Apollo 11

The NASA Mission Reports
Volume 2

Compiled from the NASA archives & Edited
by Robert Godwin

Special thanks to Buzz Aldrin, Frank Sietzen
Mark Kahn and Andy Chaikin

All rights reserved under article two of the Berne Copyright Convention (1971).
We acknowledge the financial support of the Government of Canada through the
Book Publishing Industry Development Program for our publishing activities.
Published by Apogee Books an imprint of Collector s Guide Publishing Inc., Box 62034, Burlington, Ontario, Canada, L7R 4K2
Printed and bound in Canada by Webcom Ltd of Toronto
Apollo 11   The NASA Mission Reports Volume 2
Edited by Robert Godwin
ISBN 1-896522-49-1
' 1999
All photos courtesy of NASA

INTRODUCTION

After the successful conclusion of the flight of Apollo 11 people around the world attempted to determine the significance of this remarkable achievement. Inevitably the lines were clearly drawn between those who applauded the technical bravado and those who contemplated the sociological impact.

Many people saw July 20th 1969 as the first step towards man s domination of space a concrete foundation on which the human species would build enormous space-stations, cities on the moon and ultimately star-ships to navigate the depths of interstellar space. Then there were those whose vision (or lack thereof) was more firmly rooted on Earth. They rejected the entire concept of the exploration of the heavens as humanity s ultimate destiny; offering up dubious schemes that showed the cancellation of Apollo would somehow lead to utopia here on Earth.

Few today would dispute the technological gains made by Apollo, although sheer ignorance still makes its voice heard loud and clear. Despite NASA s budget being slashed to a fraction of what it was in the 60 s and the United States economy soaring to unprecedented heights, we are still in the same distopian landscape, surrounded by poverty, hunger, war and ignorance. So much for *that* argument.

The beacon of hope that was Apollo gave the world such an unprecedented wealth of new technology that it is almost impossible to quantify it. Suffice to say that the technological darling of the moment, the Internet, would not be as it is today without the advances made in high speed data transfer that was required to fly a lunar mission.

But what of the social impact? There are still those who obstruct the next obvious step forward perhaps to Mars, or back to the moon permanently. They do so because it is easy to compare the price of a Saturn V with the cost of a new school or meals in a poverty stricken area. NASA s budget is open for all to see and is thus an easy target. However, the inexorable momentum that was initiated by Apollo will soon become evident when private corporations begin to venture into space. At that point the trade-off between an Earth-bound utopia and space exploration will become moot. The visionary entrepreneurs of the future will not have to answer to disgruntled special interest groups and their efforts will ultimately do more to revitalize the social structure here on Earth than any Earth-bound lobby group.

The social ramifications of Apollo are almost as manifold as the technological. It has been pointed out that within two years of the return of that startling picture of Earth by Apollo 8 we saw the formation of the United States Environmental Protection Agency, Greenpeace, and the instigation of Earth Day. These are just a few of the profound changes made to our thinking by Apollo.

Visionary scientist Dr Robert Zubrin (who has instigated a complete re-think of NASA s plans for a manned Mars mission) is one of many who sees the stagnation of our society if we stop exploration. Indeed, if the past is anything to go by, exploration and cultural strength have always been constant companions.

These kind of profound arguments were inevitable in the wake of Apollo, but where do the men who flew the missions stand on these issues?

The Apollo astronauts were a hand-picked elite. Mostly chosen from the ranks of the American military because of their natural ability to get the job done with a minimum of fuss. They were not favored for their eloquence. Sadly this often led to them becoming easy targets for the media. Too often the hacks moaned and complained about the sterility of the conversations between the moon and Earth and then the next day complained about the use of expletives when things went awry, such as on Apollo 10. Perhaps some expected a more prosaic description of the heavens, but when the crew of Apollo 8 chose to read from the book of Genesis they were promptly condemned by many non-Christian denominations. Clearly this was a no-win scenario better suited to Captain Kirk.

On returning to Earth the Apollo astronauts had to appear at endless banquets and were expected to be witty and charming, signing autographs and posing for pictures. Most of them just wanted to return to their lives with their families and get back to working on their next assignment. Unfortunately the public expected them to become heroes and oracles. They were counted upon to have the answers to the fundamental questions. It took years before they could shake off these expectations.

Some of the Apollo veterans gradually grew into the role thrust upon them by the media. Those such as Gene Cernan and Jim Lovell are now in great demand as public speakers while others such as Neil Armstrong have vanished from the limelight preferring the quiet anonymity of a normal life. There are those such as Buzz Aldrin, Pete Conrad, Harrison Schmitt, Dave Scott and John Young who continue to vigorously promote the future of space exploration. All of these men continue to share one thing in common with the rest of us. They are still wrestling with the answers to the same questions, should we send people back to the moon and on to the planets or should we rely on the mantra of the 90 s, Been there, done that.

It is easy to see why people expected these men to have the answers, after all they were uniquely qualified by virtue of having been out there, but that was not why they were hired. They were hired because they could get the job done in an orderly and reliable fashion.

On returning to Earth, Armstrong, Aldrin and Collins were obliged to endure a rigorous quarantine procedure. This process involved them spending three weeks at the Lunar Receiving Laboratory in Houston. The mission protocols had assigned this time as the perfect opportunity to conduct a debriefing of the crew while the details were still sharp in memory. Undoubtedly there were thousands of journalists who would have loved to have been there for that session but it was closed and classified. It was a chance for the crew to reminisce about their mission in great detail without the distractions of cameras and without having to answer poorly conceived questions. Just the facts.

This is the story of Apollo 11 as told by Armstrong, Aldrin and Collins on July 31st 1969. It is unique and has never been published before. In the pages ahead you will find many obscure acronyms and technical references, and unless you are an astronaut or rocketry expert you will undoubtedly need the acronym reference chart provided to make sense of some of the remarks. The text is unchanged and suggests that the three men were working to a checklist which had been prepared to cover the flight from suit-up to arriving at the LRL in Houston. Once you get past the jargon a surprising amount of insight is revealed about the nature of the crew and of their remarkable spacecraft. It does not, however, address any of those fundamental sociological questions and is not likely to win a Pulitzer prize for its Earth-shattering revelations about man s destiny in space. It is a transcript of three men talking about their job. What it was like to fly to the moon. For many that is historic enough.

Robert Godwin
(Editor)

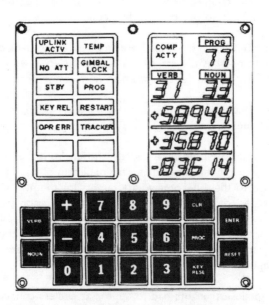

The Apollo Guidance Computer DSKY Interface

The Apollo 11
Technical Crew Debriefing
July 31st 1969
Prepared by: Mission Operations Branch
Flight Crew Support Division
Volume 1 & 2

CONTENTS

(Editor s note: The pictures included in the following pages were not part of the original document. A mixture of simulation and actual flight photos have been added.)

1.0 SUITING AND INGRESS

1.1 SENSOR APPLICATION

ALDRIN   The center lead dried out in flight. I was shaved in that area, but it dried out anyway. The one on my right chest, must have interfered in some way with the suit, because when the suit was taken off, there was a small laceration on the outside toward the rear of that particular sensor. I think thats been documented in the medical examination.

1.2 SUITING

ALDRIN   We seemed to have plenty of pad in the time frame for suiting. We were sitting around suited up at least 20 minutes before moving out to the pad.

ARMSTRONG   We had a reasonable amount of pad time to handle the little problems you might have at times. The timeline on suiting was good.

1.3 LIFE SUPPORT EQUIPMENT

COLLINS   No problem with life support equipment or transportation out to the pad.

1.6 PERSONAL COMFORT

COLLINS   The only personal comfort problem I had was that my suit fit was too tight through in the region of the UCD. During CDDT, I was really very, very uncomfortable for a couple of hours with the UCD pushing into me. This problem goes back to that first suit fit at the factory. ILC is very concerned about the mobility inside a pressurized suit, and I think they went a little bit overboard in cutting that thing on the tight side. I didn t really put the UCD on; you know what I mean. They ve got a house UCD up there, and you sort of slap that inside the suit and then you get a fit check. The only time it hurt me was when I actually had the UCD securely held and I was strapped into the couch and my legs were up. The only thing I could suggest is that when anybody goes to the factory, they take their own UCD and put the damn thing on and, during that fit check, go through some kind of an imitation of the watch position with the correct leg-to-body angle which you have in that couch for launch position. Put your own UCD on and see whether that s going to be comfortable or not. I fiddled and diddled with it between CDDT and launch, and it was still fairly uncomfortable for launch; for CDDT, that damn

thing almost did me in. Don t let them cut the suit too tight, and try to get a good fit check at the factory.

1.5 ELEVATOR AND FLIGHT DECK

ALDRIN   From the center-couch position, it s a very pleasant time period because I d sit in the elevator and walk around up there on the flight deck and contemplate just about everything, including the outside world. CDDT was a very pleasurable experience, looking out over the whole beach.

1.8 INGRESS

ARMSTRONG While we were completing the countdown procedures, the number 2 rotation hand controller was raised to the launch position. At that point, it somehow managed to attach itself to the shock attenuator release on the lower left strut. It released after a good bit of work and coordination between Mike and Fred, the BCMP. It was re-locked. No new procedure there, it just requires care and properly installing those handrests to avoid a recurrence of that problem. It would be well for the BCMP to assure himself that he knows how to re-lock any one of the strut releases that might come disengaged in this time period.

COLLINS The crew should know about the strut softeners just in case one of them gets pulled loose inadvertently in flight. You should know how to reset them. This should be added to their list of things to learn.

ALDRIN I don t feel that we really need life preservers on for launch. They interfere with what little mobility you have. It appears to me that in any abort condition you don t need to make use of the life preservers and that it would be a fairly simple thing to get them out of the little pouches that are in the L-shaped bag.

1.10 COMFORT IN COUCH

ARMSTRONG Temperature was good in our spacecraft during both CDDT and launch. I didn t suffer any of the abnormally low temperature conditions that had been reported on some of the previous flights.

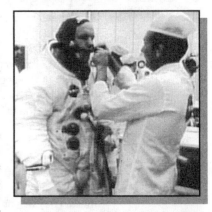

COLLINS The reason was that we were flowing glycol through the secondary loop. I believe this was the first time they tried this. The secondary glycol loop pump was on and it was flowing through the suit circuit heat exchanger. I don t know what Apollo 10 did, but I remember Apollo 9 described this deal of going bypass on the heat exchanger for 15 seconds and all that. We didn t have to mess with that at all. Our procedure worked very well. I don t know who thought it up to use the secondary loop, but it made the system very comfortable, and I recommend that they continue to do it the way we did it.

1.12 VIBRATION OR NOISE SENSATIONS

COLLINS They called out everything. Every time we were going to feel something, they were very good about calling it up.

ALDRIN We did observe some booster valving. They called it out, and it was quite obvious when there was valving taking place.

2.0 STATUS CHECKS AND COUNTDOWN

2.2 COMM VERIFICATION

ARMSTRONG   Our prelaunch COMM checks were all reasonably good.

ALDRIN   It s unfortunate that, because of the location of that center panel, we do have to split the COMM and take the center couch off the pad COMM. I can t say we really suffered much on account of it, but it would be nice if there were some way to make that switch position change - either figure out some way to loosen the belt and get back up there and readjust the COMM, or change positions in some way.

2.4 G&C VERIFICATION

ARMSTRONG   GDC align was good.

2.5 GROUND COMMAND COUNTDOWN

ARMSTRONG   Communications were excellent throughout the prelaunch phase. We had no problems with controls and displays that I can recall.

2.10 CREW STATION CONTROLS AND DISPLAYS

ALDRIN   They had that attenuation strut positioned very nicely so that I could see the altimeter. On the simulator it s very difficult from the center couch to see the altimeter. They had rotated this handle on the X strut on the left of my seat so that I could see just about the entire altimeter, which is good. I think that ought to be a standard procedure.

2.11 DISTINCTION OF SOUNDS IN THE LAUNCH VEHICLE

COLLINS   They called all those out. I thought they did an excellent job of warning us of what to expect. Not that it really makes a heck of a lot of difference because you got to sit there anyway, but it s nice to know.

2.12 VEHICLE SWAY PRIOR TO IGNITION, SWING ARM RETRACT

ALDRIN   Well, it wasn t much of a jolt when that swing arm moved out and came back in again.

ARMSTRONG   No. It was reasonably smooth. I didn t really note any vehicle sway prior to ignition.

3.0 POWERED FLIGHT

3.1 S-IC IGNITION

ALDRIN   There really wasn t much of a cue at all that I could recall. I can t remember feeling much of anything before T zero. How about the rest of you?

COLLINS   No. It was very quiet. You could feel the engines were starting up because there was a low amplitude vibration.

3.2 COMM AUDIBILITY AT IGNITION

ARMSTRONG   COMM audibility at ignition was good. Noise vibration intensity prior to release was minor.

3.3 NOISE/ VIBRATION INTENSITY SHOCK & CREW SENSATION PRIOR TO RELEASE

COLLINS   There was low noise, moderate vibration. I d say light to moderate vibration. I didn t really notice much vibration until we released. Crew sensation prior to release is just about what you d expect from Titan or from previous crew briefings on the Saturn V. It was quite mild prior to release, I thought.

3.4 HOLDDOWN RELEASE

ARMSTRONG   Now, release itself, I think we have a little bit of difference there. I felt that I could detect release, and I think your comments were that perhaps you weren t quite so sure what the moment of release was.

3.5 LIFT-OFF

ALDRIN   I can t recall any sudden change that occurred at that point, but it seems to me that there was a gradual sensation of upward movement. Then the vibration - well, it was more of an oscillation, I think, than a vibration. It certainly wasn t just longitudinal; it was a fair amount of motion in both the Y-direction and in the Z-direction. I don t know what the frequency was, but I d call it a couple of cycles per second. It was a little surprising to me, and this started rather suddenly.

COLLINS   About the time of lift-off, that s what I thought. I couldn t detect lift-off by the conventional means of sensing a transverse acceleration. However, the moment of lift-off was very apparent because this vehicle, which had been rigidly held, was now suddenly released and we were getting all manner of oscillations - X, Y, and Z, as near as I could tell. All of a sudden, this thing changed character from a static to a dynamic situation, and that was what I would call the instant of lift-off.

ARMSTRONG   Concerning   the noise/vibration intensity at lift-off, it was my impression that the combination was rather severe until approximately the time of Tower clear, at which time there was a significant decrease.

COLLINS   Yes, but would you say noise? I would say vibrations. I thought the noise level was much less than I had expected. The vibration was more.

ALDRIN   How about a rumbling? That is physically felt as much as heard.

COLLINS   You don t hear it in your ears. You feel it in your whole body. Whether

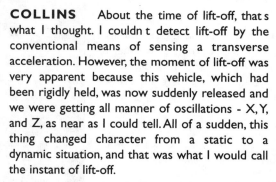

that s noise or whether that s vibration, I don t know.

ARMSTRONG   I would agree that the noise was low level.

COLLINS   In terms of interference with communications, though, I think you would also have to say that it is low level.

ARMSTRONG   That s true.

ALDRIN   Subjectively, the first 10 to 12 seconds until tower clear took longer than I thought it would. I would have thought a long 10 or 12 seconds would have been over just like that.

COLLINS   It was a long, long time before anybody gave Tower clear.

ARMSTRONG   It was right on time. I was looking at the event timer.

COLLINS   Was it 14 seconds - something like that to tower clear?

ARMSTRONG   I d say 12.

ARMSTRONG   I thought that the COMM came through quite clear.

ARMSTRONG   Instrument observation was no problem during this time. In fact, some were a lot better because the lighting in the cockpit is better than in the simulator.

ALDRIN   But we all agree that there was a decrease in the vibration, oscillation, or rumbling that could possibly be attributed to reflections off the tower.

ARMSTRONG   I think maybe it s just reflections off the ground.

COLLINS   Ground reflections.

ARMSTRONG   It goes away at about the tower-clear time.

ARMSTRONG   I thought they were.

3.13 MACH 1 AND MAX g NOISE LEVELS

ARMSTRONG   Well, I think I was surprised at how little these were apparent to me, and particularly Alpha. Alpha never came off zero throughout the launch, and I wondered if they were operating.

3.14 CONTROL RESPONSE IN HIGH g REGION

ARMSTRONG   It was as smooth as glass going through the high g region.

ALDRIN   What causes it, we don t really know, but it could be the vehicle length away from the ground; characteristic length, or whatever you call it.

3.6 LAUNCH VEHICLE LIGHTS

ARMSTRONG   Launch vehicle lights, roll program, pitch program, roll complete were on time, as were the rate changes. My impression in the seat throughout this phase, as well as the subsequent first stage, was that of going over rough railroad tracks in a train in which vibrations occur in all three axes.

COLLINS   That was a rougher ride than I expected.

ARMSTRONG   There were sharp bumps in each of the three axes periodically.

COLLINS   Yes, that s right, and the gain of the system was pretty high, also.

3.11 CABIN PRESSURE (DECREASE)

COLLINS   The valves worked as advertised and started decreasing as scheduled.

ARMSTRONG   You can hear those valves relieving, I think, from all three crew stations.

ALDRIN   While they were relieving, you couldn t hear too much else.

COLLINS   I didn t think it was that bad.

3.15 EMERGENCY DETECTION SYSTEM

ARMSTRONG   No problems.

3.16 VEHICLE RESPONSE TO GIMBALING

ARMSTRONG   The outboard engine gimbaling was not really noticed.

3.17 NOISE LEVEL VARIATION

ALDRIN   We were anticipating that, but it was just a motion as I recall. There were several little jolts to your relaxing of the four g s. That s how I recall it.

COLLINS   I would say that we were well briefed on that. I mean there weren t really any surprises.

3.22 DISTINCTION OF SOUNDS AND SENSATIONS

ARMSTRONG   There were sounds and sensations during the staging.

3.23 S-IC TWO-PLANE SEPARATION

ARMSTRONG   Skirt SEP, as I recall, was heard or felt or some observable characteristic, in addition to the light going out at the time, and I can t remember if it was a bump or a noise, but there was in addition to the fact.

ALDRIN   This would give you a clue if the lights were not working, if something had happened at that point.

3.24 S-II ENGINE IGNITION

ARMSTRONG   S-II engine ignition went smooth.

3.25 GASEOUS PRODUCTS

ALDRIN   Now, that stuff that went oozing forward.

COLLINS   That staging - well, it was just like staging on the Titan. It seemed like to me that at staging the windows lit up with yellow, almost like a flash of light.

ALDRIN   Well, let s see - S-IC. I didn t like it either, because we were tossed forward, and I couldn t look out the hatch. You re the only one that had a window at that point. I don t remember anyone saying too much about that. We ll get to that a little later on the S-IIC.

ARMSTRONG   I didn t note any.

COLLINS   Smooth - smooth as glass.

ARMSTRONG   S-II ride was the smoothest I ve ever seen.

COLLINS   It really was. It was beautiful.

ARMSTRONG   Guidance initiate was as expected.

COLLINS   Tower went as advertised.

ARMSTRONG   No Q-ball transients were noted at S-II ignition. I may have been looking at them.

ARMSTRONG   Scale change was not utilized. There were no unusual noises or vibration at this point in the flight. It was all smooth.

ARMSTRONG   The PM ratio shift was observable. You could feel g s decrease.

ARMSTRONG   Tower jettison - you could watch it go. There wasn t any question about it.

ARMSTRONG   Guidance initiate was about as expected. The S-IVB staging and engine cut-off were ...

ALDRIN   Anybody notice any exhaust coming back on the windows when the BPC went? It seemed to be a pretty clear separation.

ARMSTRONG   I didn t note any. I wasn t looking out the window at that point.

COLLINS   I was, and I didn t notice any. Those windows, 2 and 4 were clear. They didn t have any deposits on them.

COLLINS   The staging sequence is a long slow one. I m sure it was about equal to the simulated values we were used to. It seems like a long time in flight to get the S-IVB ignited. The S-IVB guidance was as expected.

ALDRIN   Any comment about the gimbal motors coming on?

ARMSTRONG   The motors were put on at 6 minutes and all came on.

COLLINS   Well, you can confirm them with the fuel cell flows, and that s not something that reaches out and grabs you. If you watched those meters carefully, you could definitely say that all four gimbal motors came on.

ALDRIN   I was looking at this sort of thing later. I found that observing them several times right at the time they were coming on, you look at the current and you see that it s a fairly small but observable change in the fuel cell current, and then just about a half second later you begin to see the rise and flow. You can catch both of them if you look at the current first and then the hydrogen and oxygen flow.

COLLINS   I just looked at hydrogen flow. They say that you have to be watching closely. If you are, you can definitely say that they all four came on.

3.42 AUXILIARY PROPULSION SYSTEM

ARMSTRONG   That was particularly noted during powered flight.

3.43 POGO OSCILLATIONS OF S-IVB

ARMSTRONG   No POGO oscillations.

ALDRIN   There s a rougher ride on the S-IVB than on the S-II.

ARMSTRONG   No doubt about it.

COLLINS   I wouldn t call it POGO, but it just wasn t as smooth.

ARMSTRONG   It was a little rattley all the time.

COLLINS   It was a lot smoother than Stafford described his ride. I think we had a different S-IVB than he had.

ALDRIN   PU shift was noticeable.

COLLINS   That was very noticeable.

ALDRIN   That was quite a jolt. About as much as one engine out.

ARMSTRONG   That s probably about right.

ALDRIN   About the same change in thrust.

3.44 SEPARATION LIGHTS

ARMSTRONG   Separation lights as advertised.

3.45 DISTINCTION OF SOUNDS & VIBRATIONS

ARMSTRONG   Sounds and vibrations we ve commented on.

3.50 COMMUNICATIONS

ARMSTRONG   Communications with the ground for the go/no-go went without a problem. There was a short time period in there when we didn t hear anything. I think we gave them a call just to make sure that we still had COMM.

ALDRIN   Following the trajectory throughout boost was quite easy with the card that we had, and I found that we were within 20 to 30 feet per second V_I, and it seems to me, 5 feet per second most of the time H-dot. Guess the altitude was a little lower, wasn t it? We might note that we did elect to have this trajectory card over part of the DSKY, which did cover up some of the status lights. The right-hand column of status lights were covered up. The ones in the LEB were observable in case any of those came on.

ARMSTRONG   Engine cut-off was smooth, and we were standing by to do a manual cut-off with the LV stage switch should cut-off not occur on time.

ALDRIN   We didn t seem to elbow each other quite as much as we had in some simulator runs. The suits are big and the elbows kind of stick out, but I didn t notice any interference with our activities.

COLLINS   The only interference I noted was that Neil s suit pocket interfered with the abort handle. He was worried about that, and I was worried about that.

ARMSTRONG   The contingency sample pocket where it was strapped on the leg was riding right against the abort handle. We adjusted that as far to the interior of the thigh as we possibly could to minimize the interference, but we still were continually concerned with the fact that we might inadvertently press that thing against the top of the abort handle.

ALDRIN   Before we go on, did you all note any numbers? I have written down here: apogee 103.9, perigee 102.1.

COLLINS   They tell me that they have better sources.

ALDRIN   I m just wondering why in our checklist we re not able to write down the CSM weight and gimbal motor numbers. We certainly ought to know what those are before flight and just confirm that those numbers have been set in.

COLLINS   I don t know why you fool with them at all. They come up to you on the first PAD prior to the first burn.

ALDRIN   Everybody in the world knows what they are, and they ought to be in the checklist.

COLLINS   I don t even know where they list them. The only other thing that I had on the launch phase was there was some peculiarity in the servicing of the oxygen quantity. Oxygen tank number I had 90 percent on my gauge, and oxygen tank number 2 had 95 percent with a 5 percent differential, and they kept talking about some mission rule which allowed a maximum of 4 percent differential. All this was a little confusing to me. It sounded as if we got shortchanged in oxygen tank number 1. I m not sure if that s true, and it even occurred to me that there might have been a slight leak in tank number 1. I m sure that there wasn t, or they wouldn t have launched us. A few words on that subject would have been nice. I think as a general rule if the loadings are not nominal, it would be nice to let the crew know that they re a little off nominal. It sounded like we launched in violation of the launch mission rule.

ARMSTRONG   Differential between oxygen tanks?

COLLINS   Yes. It was 5 percent, and it sounded like the mission rule was 4 percent MAX. And I was perfectly happy to launch with that if that was the only problem. I didn t want to bring it up on the loop and make a federal case out of it. On the other hand, it would have been nice to know.

ARMSTRONG   It only took 1 hour and 15 minutes to get through a perfectly normal launch with no problems.

COLLINS   We started late.

4.0 EARTH ORBIT AND SYSTEMS CHECKOUT

ARMSTRONG   It appeared that the platform was in reasonably good shape and its values compared favorably with the MSFN ephemeris. Everything went smoothly.

4.2 POST-INSERTION SYSTEMS CONFIGURATION AND CHECKS

COLLINS   The insertion checklist is fine, as far as I m concerned. After the insertion checklist, the items in the checklist on page L-2 and L-9 need some work to get them in the proper sequence.

ALDRIN   It s pretty hard to follow through on the time with all those things happening according to the time schedule that s on there, especially when you get down to the LEB.

COLLINS   The one who goes down to the LEB is sort of jumping from one place to another and back and forth. Some improvement could be made on the order in which those items are. I sort of got lazy and decided not to fight the checklist world and I just had my own order in which I was going to do them regardless of the order of the checklist. The follow-on crews ought to look at this section and have things rearranged to their liking for a minimum amount of moving around.

ALDRIN   For example: Step 7 on 2-9, the 20 minutes ECS post-insertion configuration, we were doing other things at that time and I don t believe that we were in position to be doing that until after we passed Canaries. Each person is sort of operating on his own. We know we re going about in the various systems checks, and that doesn t fit into a real good timeline.

COLLINS   An example here on page L2-8, item 4, EMERGENCY CABIN PRESSURE valve to BOTH. That check is made prior to anybody s going into the LEB. That s impossible to do; obviously, you have to be down in the LEB to see it. The man who goes down to the LEB - if he goes through steps sequentially as written in the checklist - would start jumping from one place to another back and forth. Some improvement could be made in the order. Now, I sort of got lazy and decided not to fight the checklist world. I just had my own order in which I was going to do them regardless of the order that is in the printed checklist. But to really be precise about it, the following crews ought to look at this little section and get things rearranged to their liking and for a minimum amount of moving around.

4.3 INITIATION OF TIME BASE 6 - AWARENESS

ARMSTRONG   Okay. Initiation of time base 6. I think we ll postpone that.

4.4 ORDEAL

ARMSTRONG   Now then, ORDEAL: We used a system where the CMP was already in the LEB and under the couches, released the latch on the ORDEAL, and let it float up to the CDR who was still strapped down in the couch that was no problem.

COLLINS   Here again, that probably should be a checklist item if people want the CMP to do that, as we did it. Then it probably ought to be written in his list of things to do.

ARMSTRONG   That worked well for us, I think.

COLLINS   Worked fine.

ALDRIN   Optics Cover Check.

COLLINS   The only thing I can say as a general rule is it goes back to this thing about becoming ill. And that is, if you re really worried about anybody becoming ill, the guy you re going to worry about is the one who s rattling around down in the LEB. In our flight, that was I and I was also the one who would be doing the transposition and docking. So I was trying to move around with minimum head movements and go minimum distances and so forth. But on the other hand, if you re convinced you re not going to be sick, well then, all those things go away. It s sort of a nebulous area. I don t know what to do about it.

ALDRIN   Well, it s something you can t afford to get ahead of yourself and be moving around too fast. If there s any question at all and I think we all played it very cautiously until each of us in our own particular way realized that it was just no problem. As we adapted to it, we could go about any kind of movement that we wanted without any particular concern, But the stakes are pretty high and you can t afford to let these things get the best of you.

4.5 OPTICS COVER JETTISON (DEBRIS)

COLLINS   I heard a little noise, but I saw no debris and I could not verify that they had jettisoned. I looked through both instruments and I couldn t see that they had jettisoned.

4.6 SCS ATTITUDE REFERENCE COMPARISON

COLLINS   Okay

ARMSTRONG   It went well. No problems.

4.7 SM AND CM RCS

ARMSTRONG   We did hot RCS checks on the service module RCS prior to TLI. The intent here was to assure ourselves that we did, in fact, have an operable control system and that our hand controllers could, in fact, talk to something before committing ourselves to a lunar trajectory. We did that in MINIMUM IMPULSE and it was extremely difficult to hear the thrusters firing. It was impossible to read an effect on any indicator in the cockpit.

COLLINS   This is with helmet and gloves on.

ARMSTRONG   Helmet and gloves on. So we were pleased when the ground said they could, in fact, see the thrusters firing. We did have to repeat one which they didn t see.

ALDRIN   I don t recall why we had the helmets on at that point.

ARMSTRONG   We didn t take them off?

COLLINS   We took them off and we put them back on.

ALDRIN   That should have been with at least one man with his helmet off so he could hear it.

ARMSTRONG   Right.

COLLINS   Well, on the other hand, if you scheduled it over the States and the ground verifies it, you don t much care.

ARMSTRONG   I was satisfied that we did, in fact, prove the point that we wanted to prove.

4.8 COAS INSTALLATION AND HORIZON CHECK

ARMSTRONG   Unstowage; COAS installation: I don t recall any points there.

4.9 UNSTOWAGE AND CAMERA PREPARATION

ARMSTRONG   We had the TV camera preparation also in the same time period; any comments there?

COLLINS   Well, again, this camera preparation probably should be written into the checklist on page L2-9, in a bit more detail than it is.

ALDRIN   Well, on 2-13 in detail, but do you want it sooner?

COLLINS   Well, this is when you re unstowing it, because really all it says is cameras and that really means the 16-mm plus the 70-mm and the various lenses. You hand them up and you get the bracket from Neil. It s really sort of an assembly process there. This is sort of a dealer s choice, but I suggest that the following crews give some thought right on page 2-9 to deciding what cameras they want to unstow or what they want to do with them and how they re going to do it. Otherwise, they re going to have another trip back down to the LEB which really isn t necessary.

4.10 DOCKING PROBE

ARMSTRONG   I think they re attempting to recall anything that might have gone abnormally.

COLLINS   There were no findings in that docking probe.

4.12 COMMUNICATIONS

ARMSTRONG   Communications were more or less in and out in earth orbit. Sometimes there was quite good S-band through the various stations; other times, it was only medium.

4.14 COMMENTS ON EARTH-ORBITAL OPERATIONS

ARMSTRONG   Any comments on the earth-orbit operations?

COLLINS   I think, in general, that s a very nice timeline. We hammered away at it enough to where we re only checking those things that really should be checked and there s plenty of time available to check them in a leisurely fashion. I think that s a nice timeline.

ALDRIN   How about the alignment? On the Saturn?

COLLINS   The platform alignment?

ALDRIN   Yes, torquing angle we got from the ground and the alignment they gave us.

COLLINS   Yes, I didn t know what to say about that. I think that s probably within normal tolerance.

ALDRIN   Yes.

COLLINS   The alignment at ORB RATE is no problem as other flights have reported.

4.18 EMS DELTA V

ARMSTRONG   Okay. EMS DELTA V. No problems.

4.19 SCALE CHANGE

ARMSTRONG   Scale change, systems, engine alignments, GO for TLI, and then .

COLLINS   Glad to get it - no problems.

ALDRIN   Well, I think it s worthy to note that we did intend to have the TV camera out. It did not seem to crowd the timeline, trying to get those pictures coming up on the West Coast. We still seemed to have a very comfortable approach to TLI.

COLLINS   That s right. Of course, that s where we said we weren t going to fool with the television if we were rushed or behind the timeline.

4.23 DRIFT TEST

ARMSTRONG   The drift test has to do with your alignment, I guess and also with the GDC drift, which was acceptable.

4.24 CREW READINESS AND COMFORT

ARMSTRONG   I think we were ready for TLI. We were unrushed and had no problems there.

4.25 SUBJECTIVE REACTION ON WEIGHTLESSNESS

ARMSTRONG   Well, perhaps a little bit of fullness in the head.

ALDRIN   I didn t notice that quite as soon coming on as in Gemini.

ARMSTRONG   Yes, I didn t feel that it was as marked as I had remembered.

COLLINS   It s so slight, that if you have anything else in your mind, you just try to ignore it. I mean, it s not any big effect.

ALDRIN   Well, there s the feeling that your face tends to lift up a little bit.

COLLINS   Yes, it does. Your eyes are puffy.

ARMSTRONG   A sensation of head-down position. I guess I had that sensation and expected it and thought it ought to be there, because we were head-down.

ARMSTRONG   Vertigo spatial disorientation.

ALDRIN   No problems.

COLLINS   None.

ALDRIN   As far as I was concerned, there wasn t anything really to be alarmed about in the least. I do think that the fact that you ve been through it makes a good bit of difference. There was a good bit made of this sort of thing before the flight, and I think someone who had not flown before would have been a little bit overly concerned.

ARMSTRONG   Yes, we were probably a little bit overly apprehensive about this area, because there had been so many comments on it in recent flights; we just didn t run into any problems.

5.0 TLI THROUGH S-IVB CLOSEOUT

5.1 TLI BURN MONITOR PROCEDURE

ARMSTRONG   The procedure went very well

COLLINS   Except yaw.

ARMSTRONG   No, the yaw was perfectly on, but the pitch showed approximately a 1-1/2-degree bias from the value that we would have expected. That is to say, with the ORDEAL set in a LUNAR/ 200 configuration, and being at the proper point on the minute each minute prior to ignition. The pitch attitude was indicating about 1-1/2 degrees higher, that is, 1-1/2 degrees to the right or plus 1-1/2 degrees from zero. We expected approximately zero. I think this would be wise to look at that carefully with DCPS training guide with respect to the adequacy of that procedure and see where that little bit of difference occurs. Other than that, the TLI monitoring went just as expected.

ALDRIN   But that was an instrument that was used to make changes if we were in control. The closing of the loop was really the observation of the H-dot which was surprisingly close. At each 30-second period, we closed the DSKY and looked at the H-dot and it was amazingly close. Of course, there get to be some pretty good H-dots at about 4 minutes and 30 seconds at about 2200 H-dot; and I don t think it was off more than 10 ft/sec at that point, so much closer than we ve seen in any simulations, right in the groove.

ARMSTRONG   Had we gone to manual TLI, then we would have probably been a little bit off in pitch. I think we had soon seen that our H-dot was beginning to get out of bounds and we made a correction, but we should understand that a little better.

ARMSTRONG   S-IVB performance and engine cut-off were outstanding.

ALDRIN   The time of the burn.

ARMSTRONG   Burn time was not quite book value, there. Did we write that down in our checklist? Burn time? Give them a burn status report?

ALDRIN   As I recall, it was a little longer than normal.

ARMSTRONG   No, as I remember, it was like a couple of seconds off in burn time, but I just don t recall now what the difference was; but other than that, it went very well.

ALDRIN   Let me just note some numbers here that was recorded at freezing the DSKY after cut-off and you are bound to miss that by a couple of seconds. The expected VI was 35575 and I reported 35579; the H-dot expected was 4285 and I have 4321; and, of course, H-dot was building fairly rapidly and that s not quite a mile a second, so the expected altitude was 174 and we read 176. The EMS was 3.3 plus.

ARMSTRONG   Yes, we knew, when the EMS showed only 3 ft/sec off in a 10,000-ft/sec burn, it was going to be pretty good to us.

ARMSTRONG   No problem.

ALDRIN   It was right on schedule and no comment.

ALDRIN   No problems.

ALDRIN   Well how about that high O2 flow anomaly that I think the ground picked up?

ARMSTRONG   Yes, I guess that s right.

ARMSTRONG   Good.

ARMSTRONG   Good.

ARMSTRONG   All on time.

ARMSTRONG   No problem.

COLLINS   The only comment on separation from SLA is the general comment about the EMS during the separation, turnaround, and the docking was that the EMS numbers got confused. The EMS got jolted and did not record some acceleration that it should have or it recorded some that it should not have; I don t know which is the case. I used the EMS as an indicator after turnaround as to how much DELTA-V to apply thrusting back toward the booster. When I got to that stage of the game, the EMS numbers made no sense at all. They were 1-1/2 ft/sec in error, and at docking, that situation continued. The EMS number that I jotted down at docking was 99.1. There s no way that the EMS could read 99.1 at docking. As I recall, I thrusted away from the booster until the EMS DELTA-V counter read 100.8, just like the procedures said. Then I thrusted minus X until the DELTA-V counter read minus 100.5. I think I thrusted plus X until it read minus 100.6. The point where the EMS was in error came after that. That s what I don t understand. When I completed the turnaround maneuver, the EMS should have read minus 101.1 and it didn t. It read down in the 90 s. At docking, when it should have read 101 plus, it read 99.1. So there is a funny there in the EMS.

ARMSTRONG   That took us quite a while.

COLLINS   You did that later.

ALDRIN   Yes, we didn t get that done until after docking.

COLLINS   Transposition and docking, in general, worked in flight just as it worked a couple of times in the simulator. I went MANUAL ATTITUDE PITCH to ACCEL COMMAND, and I started to pitch up. After 10 or 20 degrees of pitch-up, when it was definitely established that the attitude error needle in pitch was full scale high (indicating that the DAP wished to continue the maneuver in the same direction in which I had started it), then I went PROCEED and MANUAL ATTITUDE PITCH to RATE COMMAND. Then, just as in the simulator, the DAP rolled itself out. It ceased its pitch rate. I don t understand that. At the time, Buzz said that I had forgotten one PROCEED. As I recall, I went through this turnaround procedure exactly as the checklist was written. In the simulator, sometimes it worked like magic and other times it wouldn t. In flight, it worked just exactly like a bad simulator did. MIT or G&C people should check and see what if anything is wrong with this procedure. If I were going to fly this flight over again, I would say it doesn t matter if you pitch up or down. You ought to put those NOUN 22 values in there, hit PROCEED twice, and let the spacecraft turn itself around. You re going to get around within 30 or 45 seconds anyway. It s such a neat, simple, clean, easy procedure to do that way. The way we ve got it designed, to make sure that we go pitch-up instead of pitch down, sort of mixes apples and oranges. Let the DAP do it, then you take control away from the DAP, then you give it back to the DAP; and, for reasons unknown to me, sometimes it works and sometimes it doesn t.

ARMSTRONG   I d say that the manual procedure is probably the best. That would be my preference.

COLLINS   This is something that I m sure Apollo 12 and other flights will want to massage. I m firmly convinced that the way to save gas on that maneuver is to let the DAP do it. Make it a totally automatic DAP maneuver. The price you pay for that is that you never know whether it s going to pitch up or down. This is not important. In an effort to save gas and to assure that we always pitched up, I ended up wasting some gas.

5.18 STABILIZATION & ALIGNMENT AT 50 FEET

COLLINS   My procedure was worked out so I d be 66 feet away from the booster at turnaround. Because of these delays and because of the fact that the DAP kept trying to stop its turnaround rate, I would say that we were about 100 feet away from the booster when I finally turned around. This cost extra gas in getting back to it. I don t know how much extra gas, they said 12 to 18 pounds over. I don t know how much they allocated. I think it was 60 or 70 pounds. That whole maneuver probably cost 80 pounds. In the simulator, doing it completed automated, I can probably do it for 30 to 35 pounds. The difference between 30 to 35 pounds and probably 80 pounds was just wasted gas.

5.19 DOCKING

COLLINS   Docking, as in the simulator, was very easy. I did have a slight roll misalignment. I knew I had a slight roll misalignment, but everything else was lined up. Rather than diddle with it and make a last-minute correction, I just accepted it. It turned out later to be 2 degrees in the tunnel.

5.20 PHOTOGRAPHY DURING TRANSPOSITION & DOCKING

ALDRIN   We used the 16-mm camera. We used the settings that were listed in the checklist. We ll just have to look at how the film turned out before we can say too much more about that. I did use a fair amount of film and I think the pictures should come out reasonably well.

5.21 CSM HANDLING CHARACTERISTICS DURING DOCKING

COLLINS   Absolutely normal. I docked in CMC, AUTO, narrow deadband with a 2-deg/sec rate. I went to CMC, FREE, at contact. Docking alignment was fine.

5.23 ADEQUACY OF SUNLIGHT

COLLINS   More than adequate. There was plenty of sunlight. CSM docking lights were not required. The COAS reticle brightness, even with that filter removed, was still quite dim at points during the docking. It is discernible if you really look closely. At the end when you need it, it s more visible than it is 20 to 30 feet out. I would say that the COAS is marginable, but satisfactory.

5.24 CABIN PURGE AND LM/CSM PRESSURE EQUALIZATION

COLLINS   I believe all that went just about exactly as per the numbers.

ALDRIN   We went PRESSURE EQUALIZATION valve to OPEN. Where it says go to A, we went to 3.8. That s where it stabilized. Repressure O2 only brought it up to 4.4. That gave us a DELTA-P of near zero. There wasn t any cycling back and forth. There was just one cycle open and that s as far as it went.

COLLINS   That cycling back and forth only applies if you have a problem when you don t have in the full volume of the LM.

5.25 CONFIGURING FOR LM EJECTION, DOCKING PROBE, VENTING LATCHES, UMBILICALS, POWER, AND TEMPERATURE

COLLINS   Okay. The only funny here was when I opened the hatch to get into the tunnel, there was a peculiar odor in the tunnel. This odor was not exactly the same as burned electrical insulation.

ALDRIN   You commented that the wiring in the cables seemed to retain this odor.

COLLINS   I think that this is just normal. Fabric will retain an odor where metal will not.

ALDRIN   I ve noticed that same odor as characteristic of some of these new materials we have. A lot of the bags, when you get them right close to you, have this same burned-insulation odor. I m not sure if that s it, but that might explain it.

COLLINS   I don t know. My first impression was that something was burning or had been burned inside that tunnel. I went over every inch of wiring and all the connectors. I got a flash light and looked at everything. It all looked absolutely normal. We chose not to discuss it with the ground because we hadn t popped any circuit breakers and everything looked normal. It seemed like evidence of a past problem rather than an existing one

ALDRIN   I think it would be a good idea for subsequent crews to sniff around and smell what this probe and umbilicals smell like beforehand.

COLLINS   They don t smell anything like that. This was a sharp odor. I mean this was enough to knock you down when you opened the tunnel. It was one strong odor.

ALDRIN   This stuff had been exposed to a vacuum.

COLLINS   It had been exposed to the boost environment, too. I don t know how stuff would get under there with the BPC on. The BPC doesn t leave until you re darn near in a vacuum. Despite that, I thought that perhaps there was some odor associated with the high temperature of boost that had somehow gotten through the BPC and through that little tunnel vent line into the tunnel area. It sure smelled, and it smelled a couple of days later coming up on LOI. When I went in to activate the LM, the odor was just as strong. All these latches made. Latch number 6, which is the one that had acted up a little bit down at the Cape during tests, was the only one that needed one actuation to cock rather than two. Other than 6, all the others said that they were going to require two pulls to cock and they did. All that hardware worked well. We followed the checklist. We extracted in CMC, FREE, and then went to DAP control and fired the aft thrusters for 3 seconds. We went to CMC and DAP control 5 seconds after spring actuations. Neil and I both read a memo put out by MPAD, saying that for some failure modes you weren t supposed to do that; instead use SCS control. Ken Mattingly and I spent a lot of time the last couple of days before the flight trying to check all that out. It turned out to be sort of a witch hunt. For future flights, they might check into which is the best control mode for extracting the LM. I think it s okay the way we did it, but if one of the springs gets hung up and throws you sideways, it may be better to do that maneuver under SCS control rather than CMC, FREE.

5.28 VEHICLE DYNAMICS OF CSM/LM DURING EJECTION FROM S-IVB

COLLINS   There were no abnormal dynamics. The thing backed out absolutely symmetrically as far as I could tell.

5.29 ADEQUACY OF ATTITUDE CONTROL AND STABILITY

COLLINS   The S-IVB was always very stable prior to, during, and after LM separation. SM RCS plumes had absolutely no effect on visibility or on S-IVB stability.

5.32 EVASIVE MANEUVERS

COLLINS   We thought at one time we might be somewhat rushed during that time period. It turned out that was comfortable and we were prepared to do the evasive maneuver. We could have done it 5 or 10 minutes earlier than it was called for. Luckily, it is not a maneuver that is time-critical. I think the present scheme of causing the S-IVB to over-burn by 2 meters per second, and then intentionally burning the SPS for 3 seconds to compensate for that over-burn appears to be a sound procedure. I recommend no changes to it.

ALDRIN   I did notice the oxidizer unbalance to start out because it was bouncing around, but I have a note down here on the evasive maneuver that it changed from minus 180 to 130 decrease. That s only 3 seconds of burn, but you could see that this thing was in its decrease position all the time, which is what we expected. We just left it alone during that short burn. We got the first gimbal motor off a little bit before I was able to confirm it, so we had to go through a little rain dance of turning that back on and then back off again. That took a little extra time, and we used up a little extra amp-hours out of the batteries, but the ground did confirm it or at least try to confirm that we did get that gimbal motor off.

5.34 S-IVB SLINGSHOT MANEUVER

COLLINS   Now, we never saw that. It seems like the attitude they gave us was not correct.

ALDRIN   It was quite a while before we picked up the S-IVB, and it was rolling with a little bit of oscillation, a little coning effect. It definitely had a good roll to it moving away. This was during the non propulsive part of the vent and you could see two streams coming out of either side as the oxidizer was vented out. Exactly opposite to each other were two cones going out. I guess the cones were 30 degrees out one side and 30 degrees out the other, so it was definitely observable.

COLLINS   But something appeared to be wrong with the attitude they gave us. I don t know whether they miscalculated or what, but they gave us an attitude to see the slingshot out the hatch window. We confirmed that we were looking through the correct window, and it wasn t there.

COLLINS   I don t know what to say about that. I guess it went normally.

ALDRIN   We didn t make use of that procedure of keeping - we didn t use it to any advantage of having the other state vector keep track of the S-IVB. It sounds Mickey Mouse, but it could have been some assistance in telling us where the S-IVB was.

ARMSTRONG   Yes.

ALDRIN   A range rate and a VERB 89. I don t know whether it s of any value; the other guys considered it anyway.

COLLINS   I think that thing of watching the S-IVB is just the dealer s choice anyway. There s no need to watch the S-IVB. It s just that if you re going to go to all the trouble of getting the ground to compute three angles with which you should be able to see the S-IVB out a certain window, then you ought to get the correct angles.

COLLINS   Nothing to say about that.

COLLINS   I don t know what to say about that.

COLLINS   In general, we got very little radiation. Of course, we were going through the Belts about this time. I don t recall that we looked at the radiation-survey meter. Did we do that? Did anybody look at that? I don t believe that was called out. We gave daily dosimeter readings, which as far as I m concerned fall in a sort of a gee-whiz category. It s just information of very little value to anybody. They have other sources for it, and I suppose it goes on somebody s graph somewhere for posterity. Other than that, I don t have anything to say about it.

ARMSTRONG   It wasn t called out.

COLLINS   And the dosimeter we just gave them a once-a-day reading on that dosimeter.

COLLINS   Just in general, I thought all these workloads and timelines were quite reasonable and had been well worked out by previous crews and I d recommend no changes to them. I thought that whole first 3 or 4 hours worth of activity was well thought out, and we were never rushed and we were never behind.

ALDRIN   Well, our positioning of different people in different seats was a little unique, so it s a little different, I think, for other flights.

COLLINS   Yes. Well, our seat position is a separate subject in itself. As far as being hurried, we were not, although the first 5 hours of the flight I thought were quite reasonable, and that s all I have to say.

6.0 TRANSLUNAR COAST

COLLINS   We realigned the IMU in Earth parking orbit. The next time we realigned it, we were, I guess, inertially fixed. I remember now that our X-torquing angle was 0.172 degrees the first time, which seems excessive to me. We asked the ground to verify and they said it seemed excessive to them and to go ahead and redo it. So I went through P52 a second time. Instead of a minus 0.172 I got a minus 0.171. The results were repeated; therefore, the ground said go ahead and torque them, and we did. I don t understand why that torquing angle was that large. I guess it was an uncompensated X-drift, which they later compensated for more accurately, because the platform was well within its limits during the remainder of the flight. Yet this does seem like a large torquing angle. Another general comment about the IMU was I couldn t get consistent star angle difference numbers. At various times in the flight, I got either 5 balls 0.01 or 0.02, and there was no correlation. As a matter of fact, there was negative correlation. The more time I took and the more precise I attempted to be, the more often I got 0.01. On a couple of marks, when I got 5 balls, I know that I was not precisely centered when I took the mark. So, I think that there was some small bias in the sextant.

6.2 DOFF PGA S

COLLINS   There were, as far as I can recall, no surprises in doffing the PGA s.

ARMSTRONG   Buzz took his off first.

ALDRIN   We were going to stow that from the back, and I was going to be the last one to put it on. Anyway, you were going to put yours on before I did.

ARMSTRONG   As a result of a day that we spent in the CMS practicing taking the suits off and stowing them in the right place, in the right order, and so on, we decided to put all suits in the L-shaped bags: Mike s in the top, Buzz s suit in the bottom section to the rear towards the upward edge or the head end of the couch, and mine in the lower part of the L-shaped bag in the lower section. That worked fine. All three suits did go in the L-shaped bag satisfactorily and could be stowed there. We left them out though for some period of time prior to stowing them to allow them to air out since they had been worn for a significant period of time prior to this. We wanted to try to dry them out before putting them in the bags for three days. That worked as planned, and we think that s a reasonable procedure.

ALDRIN   Folding them, taking a little bit of care, seemed to pay off when you got to the point of wrestling with them to stuff them in, if you did it in a somewhat methodical way like putting one arm ring inside the helmet ring, and putting the other one in the chest. I actually took all the zippers off, then folded it over the gas connectors, and then ran both legs over and around and got it as tight as possible before putting it in. Well, it went in side ways. It seems to fit into position quite well. No doubt about it; it was a bit of a wrestling match to do this and stuff it in. It just took a little bit of extra time and effort.

ARMSTRONG   Maybe we re a little over protective, but I doubt that you could really damage those airlock connectors and helmet rings and so on. It was our intent to treat those with as much caution as we could, since we were really committed to their successful operation later.

6.3 OPTICS CALIBRATION

COLLINS   Optics calibration worked all right.

6.4 PHOTOGRAPHY, EARTH AND MOON

COLLINS   We didn t photograph the Moon at this time; the Earth we did.

6.5 SYSTEM ANOMALIES

ARMSTRONG   At this time, I think, we were starting to home in on the O_2 flow discrepancies.

ALDRIN   Yes.

ARMSTRONG   I m not sure we understand it completely. The gauge was not, apparently, indicating the correct flow level and was varying with time. That s an indication for what we think might be a particular flow varying with time. It is evidenced by the fact that the quad accumulator cycle flow rate continued to decrease until it got down to about 0.3, and then it went back up to 0.4.

ALDRIN   It would register around there each time. Then it seemed to go up almost to the safe value. That led me to believe that there was nothing wrong with our gauge.

COLLINS   Sounds like the gauge was operable but out of calibration.

ALDRIN   Right.

COLLINS   We spent a lot of time with EECOM before the flight discussing what items to check and what items not to check. I suppose as long as we have space flights, we re going to have philosophical disagreements on how exhaustively we want to check all the equipment. My personal philosophy is that if you don t have some reason to believe that it s broken, leave it alone. Don t fool around with it. FOD, of course, has a number of mission rules that require verification of each and every component of each and every system to make sure that they re not going to

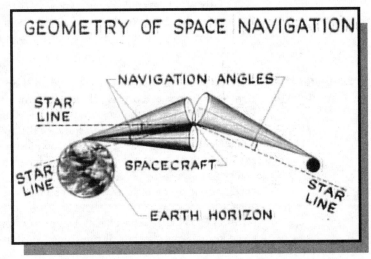

GEOMETRY OF SPACE NAVIGATION

violate one of the mission rules. I can understand their viewpoint. Maybe the truth lies somewhere in the middle. We ended up after many, many discussions including the operation to make sure that the glycol was flowing satisfactorily through the secondary radiators without any leaks and to make sure that the secondary water boiler was functioning properly. We did that pre-TLI. Then pre-LOI, we checked only for gross radiator leaks and did not check secondary water boiler operation.

ALDRIN   It wasn t really a difficult time-consuming task. It went very smoothly.

COLLINS   It s a question of whether you want to do it. For example, on the secondary glycol radiator leak check, the secondary glycol loop has been bypassed, that is, no fluid has been allowed to go through the radiator. You put the valve from bypass to flow for 30 seconds, turn the pump on, and allow fluid to flow through the radiator. Then you confirm that there is no leak by checking the accumulator quantity and making sure that it does not decrease. So what happened? In this case, accumulator quantity decreased by about 4 percent. This had never come up before. The ground suspected that it was due to thermal characteristics in contractions or expansions in the system, and not a leak. It dropped and then stabilized. I preferred to leave that equipment alone rather than mess with it. I guess there was no leak. On the other hand, we could have gotten into a big argument over a suspected leak even to the point where you might have to delay TLI by a revolution. If you don t have concrete evidence that something has malfunctioned, and it s your backup system, leave it alone. Don t mess with it.

6.7
CHLORINATE
POTABLE
WATER

ARMSTRONG   We did this once a day before bedtime. The little injector assembly got more and more difficult to operate as each day went by. The chlorine tends to stick and corrode the screw threads. What started out to be a fairly low torque application, towards the eighth day got to be a fairly difficult task to screw the container down so that the chlorine capsule in it would get squashed. We also got some leakage the first day due to the fact that I did not have the threads fully engaged. It felt to me as if I did have the threads engaged. However, when I started screwing it down, I found I didn t. Chlorine was escaping, and I had to get the towel out and mop it up. After that, I didn t have any trouble with it.

ALDRIN   I found myself invariably wanting a drink of water after we chlorinated the water. You couldn t do that unless you put some in the bag ahead of time. We should have done that. It just didn t occur to us until afterward.

COLLINS   I certainly don t think it s worth changing the system for mainline Apollo. For future spacecraft, you d certainly like some built-in way of assuring yourself of a germ-free water supply without having to go through this kind of procedure.

6.8
COMMUNICATION
SETUP FOR REST
PERIOD

ALDRIN   The way that the flight plan handled it was a little involved. We were in a translunar switch setup. It would say each time for rest period go to lunar coast except for such and such. In the LM, we had a fairly simple way of handling it. We just labeled, straight on down the line, the position of the switches. We could probably come up with something similar to this. It could include just a certain set number of switches that are all S-band. You just make a quick check of all these and have them in the right configuration, instead of having to refer back to the systems management book. Keep that checklist out of the flight plan, and keep it in the checklist.

ARMSTRONG   The checklist is pretty long, so you end up with a fairly complex piece in the flight plan and also a complex list in the systems book.

ALDRIN   But the flight plan does have two sleep mode options: high gain or OMNI. So, you really have more than you need in the flight plan.

ARMSTRONG   We insist that we only go in the OMNI mode during sleep periods. We decided that it would be best.

6.10 EASE OF
OPERATIONS OF
COMMUNICATIONS

COLLINS   They were all right. There were also times when we had communications dropout that I don t think were explained. I had the feeling that there were a lot of ground antenna switching problems. There would be times when we really should have had sound and we didn t. It was due to some sort of a ground problem. It seemed to me that there were a lot more of those problems on this flight than there were on Apollo 8.

ARMSTRONG   You probably noticed that in the Center, too, handovers and switching.

ALDRIN   We chose not to control it on board, switching from one OMNI to the other. We let the ground handle the whole thing, and they just have a choice between two OMNI s. They are going to run into some dropouts invariably.

COLLINS   The PTC rate we used was 0.3 deg/sec. For the crew to switch OMNI s manually and go around A, B, C, D during the time when they're awake is really too much of a job because you're having to switch OMNI s approximately every 5 minutes.

ALDRIN   It s 18 minutes

COLLINS   So, I think it s a correct decision to let the ground switch between opposites B and D OMNI antenna rather than having us switch manually A, B, C, D; but I guess the ground needs some refinement in that procedure because we did have a number of cases of COMM dropouts, and later on in lunar orbit, it was even more so.

6.12 PREFERRED PTC MODE & TECHNIQUE FOR INITIATION

COLLINS   There are all sorts of real varied funnies in the check list (page L9-6) for how to get into PTC. Now, just for example, during the period when you are waiting for the thruster firing activity to quiet down, there s a 20-minute nominal wait period for thruster firing to diminish. And for instance, if the crew wants to see how the thruster firing activity is coming along, the way of verification is VERB 16 NOUN 20, monitoring the gimbal angle, and watching the lack of change in the gimbal angles. Yet, if you do that and leave VERB 16 NOUN 20 displayed on the DSKY, when you proceed 8 or 10 steps later to the point where you start to spin the spacecraft up, instead of getting 0.3 deg/sec rate, you will get a rate in excess of 1 deg/sec. And this fact is not well known. This is something that we found in the simulator shortly before the flight and penciled into the checklist. But I would just say in general that that checklist should be reworked. There are many little pitfalls. For example, if you find yourself in an inertial attitude, and all you want to do is spin up around that attitude, the checklist implies that you can just go into it at that intermediate point, but that is not the case either. You must pretend that you are in the wrong attitude, ask the computer to maneuver you to the right attitude and then go through the entire checklist from that viewpoint, or it won t work properly. These are just two pitfalls that I happen to know about right now.

ARMSTRONG   It seems to me that the point is that this is a very good procedure that worked extremely well, and we re going to find that it s extremely easy to use but has not stood the test of time yet. It needs a lot more experience in use before we could use it reliably and repeatedly every time without causing a later problem that we couldn t predict.

COLLINS   That s right. Another little facet of it is that after the PTC is initiated, then there are certain no-no s in regard to the use of the DSKY s having to do with collapsing deadband and other problems internal to the computer. So, I think some explanation and expansion in those pages in the checklist is in order.

ARMSTRONG   It s probably worth noting here, while we re thinking about it, there seems to be some advantage to writing a program to do this job. At least it should be considered, rather than the one we re using at the present time. It could obviate many kinds of minor difficulties that we didn t mention until now.

6.14 EASE OF HANDLING OPTICS AND SPACECRAFT FOR NAVIGATIONAL SIGHTINGS

COLLINS   With P23, as I practiced it in the simulator and made use of the AUTO optics to maneuver the spacecraft to each star substellar point, the flaw in this technique is that the spacecraft roll angle is unconstrained in that with large trunnion angles, the computer may pick a roll attitude which causes the star to be occulted by the LM structure. Now, the flight planners came to me a couple of weeks before the flight and said that to get around this disadvantage of the AUTO optics, they wanted to use ground-computed angles to which to maneuver, and then these ground-computed angles would have a roll angle which would assure that the star would not be occulted by the LM structure. And at that time, I told them that all my training had been built toward using

AUTO optics for these maneuvers. I asked them to go back and find stars whose trunnion angles were small enough that this would not then be a problem (the LM structure occulting it). Flight planning talked to the MPAD people and said that they could not find such stars with the proper in plane/out of plane geometry. But the ground-computed angles would locate satisfactory substellar points and all subsequent maneuvers would be very small. Now, I should have called a halt right there and sat down with the flight planning people and with the MPAD people, and I should have gone through each star, each maneuver, each gimbal angle, each subsequent substellar point, and ironed out just exactly step by step how many maneuvers would be required; the size of them and exactly what was being furnished to me in regard to roll angles. However, I didn t. That s one of the things that fell through the crack. So, in flight when I maneuvered to the ground-supplied angles, I found that I was nowhere near the substellar point as determined by the fact that the sextant reticle was not parallel to the horizon at that point. And here I think we had some kind of a communications breakdown with the ground, because I kept telling them that this was not at a satisfactory substellar point, that the reticle was not parallel to the horizon. They kept telling me that it was all right to go ahead and mark anyway.

ALDRIN   They didn t really mean that. We re sure they didn t.

COLLINS   Now, I m not sure what they meant. Maybe you hit the nail on the head. What they meant was that the spacecraft did not have to be rolled in such a manner that the spacecraft roll was parallel to the substellar point. In other words, what they were saying is that the computer program could accommodate a change in

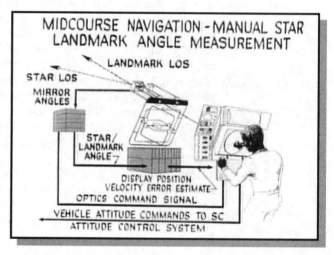

spacecraft roll simply by torquing the optics around to go off at a peculiar angle. Nonetheless, when you look through the sextant to get accurate marks, you must have the reticle pattern parallel to the horizon or you are not measuring the true angle between the star and horizon. Here s the star and the horizon, and instead of measuring this angle, you re measuring this angle or that angle or some other oblique angle that is larger than the true angle, which is the angle from the star normal to the horizon. So this initial run on P23 got very confused. The following day, the problem went away because we were far enough away from the Earth, and the fact that their angles were not at the substellar point became immaterial because the Earth was small enough that a very small maneuver on my part could locate the substellar point. But when you are close to the Earth, and the Earth is very large, and you have an obvious roll on its alignment in the reticle, then it requires a very large maneuver to maneuver the spacecraft over to the substellar point. I d be happy to draw it on a blackboard some other time for the proper people. I was reluctant to make these large maneuvers, because I thought something was wrong. And they kept saying go ahead and mark, that it was all right, and so I did take some marks and the DELTA-R s and the DELTA-V s were excessive. I don t know what else to say now. I d sort of like to get a blackboard and talk this over with flight planning and with the people from MPAD, if necessary, and see where we went wrong. It s my fault in that I didn t get all the interested parties and sit down and go through step by step and maneuver by maneuver exactly where we were going to go and what we were going to do.

ALDRIN I think it s one of these areas that it would have been nice maybe, for you anyway, to have had an abbreviated simulation with Houston as part of our training. One big problem there is that you just can t always count on the simulator giving enough fidelity.

ARMSTRONG Yes. I think that s one of the areas where the simulator probably falls a little short.

COLLINS In my mind, it s a question of time available. I had so much stuff to learn, and I had divided up the time, and P23 was a relatively small slice of the overall training. I didn t want to really spend the time to sit down and go and hammer this stuff all the way through, although it appears I should have. That s another thing. That state vector was another heartache.

ARMSTRONG The state vector may have been bad initially but especially when you get two large errors in a row. We incorporated it, and from that point on, the state vector wasn t any good.

COLLINS That s right. The state vector was mediocre to begin with and it rapidly got worse. But each star has its own distinct substellar point, and you take a measurement on two stars in a row. This requires that you maneuver from one substellar point to another. I kept telling those people that before the flight and they kept saying, Oh no. They re all right close together. I think there s some confusion on their part and maybe some on mine.

ALDRIN I think it s all unfortunate that the first mark, the first star set that we had, was changed in the flight plan.

COLLINS Well, that s another thing. We didn t mark correctly. Sometime between the last time we simulated it and the first time we pulled this out in the flight, star number 2 had been moved from the number 1 position down to number 4 position, and they had done it just by changing the 1 to a 4 and drawing a little arrow. When you read me the numbers, you didn t note that I read star number 2 and it was the same old star I had always marked on first. That was just a bad area. A little bit of work could have cleaned that up before the flight, and I just didn t have the time or the inclination to sit down and hammer it out with the people required, and I should have. Well, we were fine the next day only because the Earth was so much smaller. If you have a little Earth and you re supposed to be marking on this point and you re at this point, it s no big deal to move from here over to here. But the Earth is big and you re supposed to be marking on this point, and you re really over there; that requires a big maneuver. The same problem existed the next day. However, a tiny maneuver on my part solved the problem; whereas the day before it was a huge maneuver, and I was reluctant to make that maneuver. As a general comment, I ve found that the telescope was a very poor optical instrument in that it required long, long periods of dark adaptation before any star patterns were visible. In most cases, it was not convenient to stop and spend the amount of time necessary to make any use of the telescope. Thus, we kept our platform powered up continually. My procedure was to ignore the telescope and to take at face value what the sextant said. In other words, if the sextant AUTO optics came up with a star in the sextant field of view, I accepted it as a matter of fact that it was the correct star. We marked on that star without any further verification. I suppose this could rise up and bite you, but I felt safe and comfortable with it, and it worked throughout the flight.

COLLINS Now by that I assume they mean the ground-supplied sequence, and that I felt was fine. Got any comment about that?

ARMSTRONG Well, they may also be referring to P30 and P40 sequence and so on. And it was our intention to do those very carefully in just the way that they are detailed in the procedures; not because the burn was all that important and we compensated for it if we made an error, but rather because the analysis of that burn on the ground was going to be the thing that determined that we have a good SPS for LOI. Because that was the case, we wanted the ground not to be at all confused about what procedures we would use and just how the burn was made. So we tried to stick precisely with the same procedures you d use for an SPS burn.

COLLINS In general, I thought all the P30 s and P40 s worked out very smoothly.

ARMSTRONG Well, the first midcourse was cancelled to allow the DELTA-V value to grow in size so that the second mid course correction would be reasonably longer, allowing engine operation to be well stabilized and more accurately analyzed on the ground.

ALDRIN Midcourse 2 was 21.3 feet per second.

ARMSTRONG The results of that were very, very good and the residuals were very small, 0.30 and 0.20. But there was some question about the fact we had a relatively large EMS residual; namely, 3.8 ft/sec in about a 20-ft/sec burn. The predicted knowledge of tail-offs apparently was badly in error or else the knowledge of the EMS itself in the tail-off region was badly in error. That never was corrected throughout the flight. We saw this condition through the rest of the SPS burns.

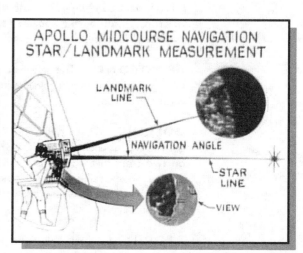

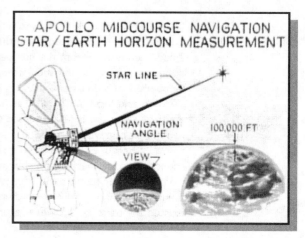

ALDRIN Did it say anything about the sextant star check? They updated that. It was pretty much out of sequence.

ARMSTRONG The first one they gave us. Then the second one was in that direction because, of course, the LM was there.

COLLINS It was our desire that insofar as possible an inertial attitude check be made (in the absence of the burn) so that if you made the burn they knew you were in fact pointing in the right inertial direction. Of course, the LM is out in front of you and you can t look down the X-axis of the optics, so you re constrained not to point any closer to the X-axis than the LM will allow. However, the initial values that they gave us were sort of like down the Z-axis. Of course, you could have the optics pointing down the Z-axis and then you could be free to pass that test and still have the spacecraft pointing 180 degrees out from where you want it to be. It will still pass that test, so in our view that wasn t particularly good. You were really just checking your alignment of the platform, which is really not what you re trying to do. You re trying to check that the

spacecraft is pointed the way you want it pointed so that was the reason for our request for additional star checks.

6.17 ADEQUACY OF CSM/MSFN COMM PERFORMANCE & PROCEDURES FOR COAST DURING AGA REFLECTIVITY TEST

COLLINS   Okay. Adequacy of all this stuff for the AGA reflectivity test. I understand we didn t have that and we cancelled that.

6.18 TELEVISION PREPARATION AND OPERATION

COLLINS   I thought in general the onboard color television system was well designed and was easy to operate. Buzz, you got anything to say about that?

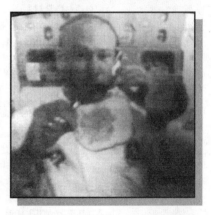

ALDRIN   It was quite easy to hook up and put together. We ended up putting the two together making use of tape instead of the Velcro that was on there, to get the monitor right close to the camera. I think initially we were a little tangled up in wires. There were wires all over the place, and we were running around from one strut to the other. We found out that it was set to have the monitor attached right to the camera itself, so from that point on, we taped the monitor beside the camera.

ARMSTRONG   Well, we have a couple of comments we ll get into sometime later with respect to television, but with respect to its operation, it s unquestionably a magnificent little piece of equipment. However, you cannot operate it without any planning at all. You do have to think about whether the vehicle is rotating or not, in what area you re going to take pictures, where the lighting is going to be from, and through what windows, and all that sort of thing. This takes some planning to enable you to assure yourself that you are going to get a good TV picture of whatever you decided you are going to take a picture of.

COLLINS   That s right, and the monkey is on the back of the crew, functioning as script writer, producer, and actor, for the daily television shows. We had no time nor inclination preflight to plan these things out so they were all sort of spur-of-the-moment shows. And maybe that s a good way to do business and maybe that s not. I don t know.

Maybe other flights with perhaps more time to devote to this should give some thought to what has previously been done and what are the best things to cover and when is the best time to present them. The next crew should spend a simulator session working out things like angles and light and what have you.

ALDRIN   There is no doubt that you want to do it right, because there s a big audience looking on.

ARMSTRONG   It inspires you a little bit when all of a sudden you have about 10 minutes left to go for a scheduled TV broadcast and the ground says there are 200 million people waiting to see you. They re all watching. Now what are you going to be showing?

COLLINS   We re trying to paint the picture of having this highly trained professional crew performing like amateurs. They don t know where to place the camera or what to do or what to say. It hasn t been well worked out. I feel uncomfortable about this.

ARMSTRONG   It s just fortunate that the camera is as good as it is and it compensates for the inabilities of the operators.

ALDRIN   I think that some of the better things that we did were just monitoring and just trying things out before we got to the point of putting on the show. I think there is the ability of people on the ground to see what s coming across, look at it, select what they want, and then assemble it together and release it. I m sure everyone wants to have a real-time picture and voice along with it, but you re going to suffer somewhat in the quality you get. For example, activity in the LM, when we were just trying to see how it was working. All of a sudden we found that we were going out live and we were completely happy with that. This was one of the better shows we did.

ARMSTRONG   I agree with that, but on the other hand there is another side to that discussion that doesn t involve somebody thinking about how that situation can be handled. We can put out something that the agency is willing to stand behind and can be proud of without the crew having to make a lot of last-minute quick guesses as to what they ought to be doing.

6.19 HIGH GAIN ANTENNA PERFORMANCE

COLLINS   It was okay I guess.

ALDRIN   It seemed to work fine. I placed it in AUTO, threw the switch over to MEDIUM or NARROW, and just a couple of seconds later the signals transferred.

ARMSTRONG   There was one observation here that seemed to me to be different from the simulator. In the spacecraft, I could seldom if ever detect a difference in signal strength between MEDIUM and NARROW. In the simulator, it s always decidedly different. The conclusion to be reached is that either the simulator is not an accurate representation of signal strength or that we really weren t getting any difference between MEDIUM and NARROW beam. We were, in fact, stuck in one or the other irrespective of switching.

ALDRIN   I would expect there wasn t as much difference between WIDE and MEDIUM, but when you went to the NARROW, you could see it. It wasn t consistent. In any case, it was unlike what we were used to and as long as the signal was received, I guess it s not a problem.

COLLINS   I think that s a function of distance, too. Now in lunar orbit, there was a noticeable difference between MEDIUM and NARROW. But there were some funnies in that high gain antenna. We were playing with it some time and we didn t have control over it and the ground had one of the OMNI s selected. We thought we were controlling it and we weren t. Another funny was that there were ground switching problems where the thing was not receiving a proper signal.

ALDRIN   I remember one time the ground said go ahead and turn the high gain off. I complied and we lost COMM. I don t think they expected it; the next time they had control, we were on OMNI at that time. It wasn t at all clear to me at all times who had control and who was running the show.

COLLINS   That s right and it was a great temptation to go to command reset to make sure that we had control, except that that threw six or eight other switches that we were reluctant to change. I suspected at times that it was not working properly. I never absolutely caught it malfunctioning. I think those suspicions mostly had to do with the fact that we didn t have control of it or the ground had some sort of a sighting problem.

ARMSTRONG   The confusion in my mind often was that I wasn t really sure what our configuration actually was. You can t tell by the switches and trying to interpret what you see in terms of displays you have available and what you hear.

6.24 RADIATORS

COLLINS   We never flowed through the secondary radiators because the primary worked fine. The cabin temperature (translunar) was slightly warmer than we would like it, although the gauge readings were quite cool. We were running 60 degrees cabin temperature and 57 degrees suit temperature.

ARMSTRONG   High 40 s in the suit and low 60 s in the cabin.

COLLINS   Yet we were warm in spite of those low numbers.

6.25 CM/LM DELTA PRESSURE

COLLINS   Well, the LM pressure would slowly decay, but remain well within tolerance. I don t have any good numbers. It was a tight LM.

6.26 RE-ESTABLISHING PTC

COLLINS   We ve already discussed that, I think. We always used 0.3 deg/sec roll and we never tried the 0.1. It would be advantageous, in regard to antenna switching if stability is satisfactory, and 0.1 deg/sec would probably be a better mode than 0.3. It would also save some gas. However, we did not investigate that. Perhaps that ought to be something for future flights to look into. I think that theory has been mentioned to FOD.

ALDRIN   Maybe.

COLLINS   We did it coming back.

ALDRIN   When it worked, it worked like a charm. There were a couple of times when it didn t seem to want to work.

COLLINS   Optics CAL the next day worked fine.

COLLINS   The fuel cells performed perfectly. Purging didn t present any problems. We followed the checklist on the heaters and they worked normally.

COLLINS   LM and tunnel pressures were normal.

COLLINS   Latches, as I say, were all verified. Latch number 6 required one actuation to cock. That was the only anomaly and it was within the realm of normal.

COLLINS   I m not sure what that means, but everything in the tunnel was normal.

COLLINS   Probe and drogue removal was absolutely normal. Have you anything to say about that?

ALDRIN   Well, as far as I m concerned there was no disorientation in going from one spacecraft to another. It was quite easy to go from one to the other. It would take a little readjusting to get yourself into position when you first entered one vehicle or the other. You weren t sure what you were looking at. But there was no disorientation associated with that.

ARMSTRONG   I didn t observe any problems with that.

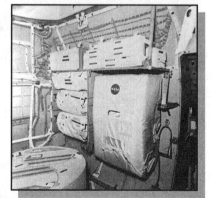

ALDRIN   We may not get back to this again, but I think that the exercise we had in the LM was extremely valuable from our standpoint. It was conducted from a very comfortable timeline. We had no particular schedule to meet; we used the camera to document. In addition, the television set at this time was quite valuable.

COLLINS   From the CMP position, it was of great value to have a one-day head start on the removal of the probe and the drogue. If problems arose with the probe or the drogue, you have time to troubleshoot with the ground. I was glad to get that probe and drogue out a day early.

ARMSTRONG   It was something you know that hadn t previously been done quite this way. It just seemed that it would make us more comfortable, going back and forth to the LM, that if there was anything wrong, we d have some chance to talk and think about it and give the ground some time to think about it. That didn t turn out to be necessary because it was perfect, but still I think all of us felt a lot more comfortable having spent some time going back and forth and checking the stowage and looking over everything. The repetition just took the pressure off the next day s IVT.

ALDRIN   Working in the very relaxed environment of the constant wear garment, there were no problems. We didn t really need to be restrained. I used the restraints and all it seemed to do was pull my pants down. You did have to be a little concerned about floating away from what you were doing, however, it was no great problem to push yourself back down to where you wanted to be.

ARMSTRONG   This enabled us to get a little ahead in stowage.

ALDRIN   Concerning transfer items: we brought several books back - updates and a couple of procedures.

ARMSTRONG   So, all in all, I guess it worked out well. We recommend it as a useful procedure.

6.38 EATING PERIODS

COLLINS   They were well spaced and I thought adequate time was given to eating. Quality of the food will be discussed later.

6.39 WORKLOADS

ARMSTRONG   The workload during the translunar coast is very light as it should be.

ALDRIN   In comparison to the preflight workload, it gave us a couple of days to relax. I think it s important to store up the rest.

COLLINS   I think so too.

6.40 REST PERIODS

ALDRIN   We re all good sleepers. The first one was not as good as the second or third, but the first sleep period was still surprisingly restful as far as I m concerned.

COLLINS   I think particularly when you get into the later flights of extended EVA s and lunar activity, somehow the crew must place themselves in a frame of mind of looking on the separation of the LM as the beginning of the flight plan and to relax, get plenty of sleep, and conserve their energies in all the events leading up to that point. To arrive in lunar orbit tired can create problems and it s possible to do that if you don t approach it in the right frame of mind.

ARMSTRONG   I think Mike s hit the nail on the head. We did precisely that. We got a lot of rest and got into lunar orbit eager to go to work and that s a particularly fortunate position to be in.

COLLINS   This is something we ve talked about before the flight and I don t know how you can get yourself in that frame of mind but I think it is a frame of mind. You have to get yourself convinced that there will be a nice relaxing couple of days going to the moon.

ALDRIN   The first unusual thing that we saw I guess was I day out or something pretty close to the moon. It had a sizeable dimension to it, so we put the monocular on it.

COLLINS   How d we see this thing? Did we just look out the window and there it was?

ALDRIN   Yes, and we weren t sure but that it might be the S-IVB. We called the ground and were told the S-IVB was 6000 miles away. We had a problem with the high gain about this time, didn t we?

COLLINS   There was something. We felt a bump or maybe I just imagined it.

ARMSTRONG   He was wondering whether the MESA had come off.

COLLINS   I don t guess we felt anything.

ALDRIN   Of course, we were seeing all sorts of little objects going by at the various dumps and then we happened to see this one brighter object going by. We couldn t think of anything else it could be other than the S-IVB. We looked at it through the monocular and it seemed to have a bit of an L shape to it.

ARMSTRONG   Like an open suitcase.

ALDRIN   We were in PTC at the time so each one of us had a chance to take a look at this and it certainly seemed to be within our vicinity and of a very sizeable dimension.

ARMSTRONG   We should say that it was right at the limit of the resolution of the eye. It was very difficult to tell just what shape it was. And there was no way to tell the size without knowing the range or the range without knowing the size.

ALDRIN   So then I got down in the LEB and started looking for it in the optics. We were grossly misled because with the sextant off focus what we saw appeared to be a cylinder.

ARMSTRONG   Or really two rings.

ALDRIN   Yes.

ARMSTRONG   Two rings. Two connected rings.

COLLINS   No, it looked like a hollow cylinder to me. It didn t look like two connected rings. You could see this thing tumbling and, when it came around end-on, you could look right down in its guts. It was a hollow cylinder. But then you could change the focus on the sextant and it would be replaced by this open-book shape. It was really weird.

ALDRIN   I guess there s not too much more to say about it other than it wasn t a cylinder.

COLLINS   It was during the period when we thought it was a cylinder that we inquired about the S-IVB and we d almost convinced ourselves that s what it had to be. But we don t have any more conclusions than that really. The fact that we didn t see it much past this one time period - we really don t have a conclusion as to what it might have been, how big it was, or how far away it was. It was something that wasn t part of the urine dump, we re pretty sure of that. Skipping ahead a bit, when we jettisoned the LM, you know we fired an explosive charge and got rid of the docking rings and the LM went boom. Pieces came off the LM. It could have been some Mylar or something that had somehow come loose from the LM.

ALDRIN We thought it could have been a panel, but it didn t appear to have that shape at all.

COLLINS That s right, and for some reason, we thought it might have been a part of the high gain antenna. It might have been about the time we had high gain antenna problems. In the back of my mind, I have some reason to suspect that its origin was from the spacecraft.

ALDRIN The other observation that I made accumulated gradually. I don t know whether I saw it the first night, but I m sure I saw it the second night. I was trying to go to sleep with all the lights out. I observed what I thought were little flashes inside the cabin, spaced a couple of minutes apart and I didn t think too much about it other than just a note in my mind that they continued to be there. I couldn t explain why my eye would see these flashes. During trans-earth coast, we had more time and I devoted more opportunity to investigating what this could have been. It was at that point that I was able to observe on two different occasions that, instead of observing just one flash, I could see double flashes, at points separated by maybe a foot. At other times, I could see a line with no direction of motion and the only thing that comes to my mind is that this is some sort of penetration. At least that s my guess, without much to support it; some penetration of some object into the spacecraft that causes an emission as it enters the cabin itself. Sometimes it was one flash on entering. Possibly departing from an entirely different part of the cabin, outside the field of view. The double flashes appeared to have an entry and then impact on something such as the struts. For a while, I thought it might have been some static electricity because I was also able, in moving my hand up and down the sleep restraint, to generate very small sparks of static electricity. But there was a definite difference between the two as I observed it more and more. I tried to correlate this with the direction of the sun. When you put the window shades up there is still a small amount of leakage. You can generally tell within 20 or 30 degrees the direction of the sun. It seemed as though they were coming from that general direction; however, I really couldn t say if there was near enough evidence to support that these things were observable on the side of the spacecraft where the sun was. A little bit of evidence seemed to support this. I asked the others if they had seen any of these and, until about the last day, they hadn t.

ARMSTRONG Buzz, I d seen some light, but I just always attributed this to sunlight, because the window covers leak a little bit of light no matter how tightly secured. The only time I observed it was the last night when we really looked for it. I spent probably an hour carefully watching the inside of the spacecraft and I probably made 50 significant observations in this period

ALDRIN Sometimes a minute or two would go by and then you d see two within the space of 10 seconds. On an average, I d say just as a guess it was maybe something like one a minute. Certainly more than enough to convince you that it wasn t an optical illusion. It did give you a rather funny feeling to contemplate that something was zapping through the cabin. There wasn t anything you could do about it.

ARMSTRONG   It could be something like Buzz suggested. Mainly a neutron or some kind of an atomic particle that would be in the visible spectrum.

7.0 LOI THROUGH LUNAR MODULE ACTIVATION

ARMSTRONG   With respect to preparation for LOI, our flight plan was written in such a way that it depended on doing mid course 4 and option I P52 to get the landing site REFSMMAT into the computer and then an option 3 REFSMMAT P52.

ALDRIN   Was that before midcourse 4 was performed?

COLLINS   Yes, midcourse 4 was with the landing site REFSMMAT.

ARMSTRONG   Then we did our simulation of LOI where we checked the gimbal motors and a 360 degree pitch maneuver to look at the Moon, followed by preparation for LOI. The midcourse 4 was cancelled. We did not do the option I P52 that established our new REFSMMAT. ... set up the computer for the LOI. When we got around to the P52 in the flight plan, which occurred at 73 hours, we did option 3. We recognized that we had never done a new P52 to an option I. We are not sure that we could at that point in time.

COLLINS   Did they have an uplink?

ARMSTRONG   I m not sure they had uplinked the necessary data into the computer. In any case, we recognized that we were not operating the way the flight plan had intended, due to this cancellation of midcourse 4; therefore, we got that information from the ground. We did a P52 option I, a P52 option 3, and our simulation of LOI where we brought the gimbal motors on and checked that everything was really copacetic. During this process we got behind the timeline because we did things differently than we had intended in the flight plan. Consequently, we cancelled the 360 degree pitch maneuver to photograph the Moon. We did not feel very bad about that since shortly before, when we went into the Moon shadow, we did look at it extensively through the windows and took a lot of pictures with the high-speed black and white film. I think we accomplished what we wanted to do in looking at the Moon from a relatively close range. We agreed to cancel the 360 degree pitch maneuver. We were then slightly ahead of the timeline in preparation for LOI. We spent a little more time discussing that among ourselves than we had planned, since it was different than our simulations.

ALDRIN   There was something else. Was it just the two different alignments that got us a little bit behind?

COLLINS   I think it was not having a REFSMMAT.

ARMSTRONG   There was something else. I do not recall right now what it might have been. We did that secondary loop check, and a secondary radiator flow check.

COLLINS   We could not see the stars. Was there a star check at a certain time? We were sitting around on one foot and then the other waiting for something. There was a time in the pad when the star check was only valid after I I past the hour.

ALDRIN   That appears at some time. I don t see that written on this particular set up.

ARMSTRONG I might mention on the sextant star checks that, on most occasions, we manually drove the optics CDUs to the ground-computed values for the star and checked the attitude in that manner. That always worked for us. We were always able to see the star in the sextant field of view by manually guiding the optics rather than using the computer to designate the optics.

7.3 SPS BURN FOR LOI-1

ARMSTRONG Now we will go up to LOI. LOI was on time, and the residuals were very low. Again we saw a large value of DELTAV_Cs - 6.8. Buzz will now comment on the PUGS.

ALDRIN We had been briefed on the experiences that Apollo 10 had had with the operation of the PUGS oxidizer blow valve, whereby they had responded to the initial decrease that the system gave them by placing the switch to DECREASE. Subsequently, it went to INCREASE. They followed it but were never able to catch up with it. It was suggested to us that the best procedure was to monitor this in the first 25 seconds, again expect it to be in DECREASE, and then expect that maybe even by going to FULL INCREASE you could not keep up with the system. With this in mind, I watched it throughout the burn. As soon as it started toward the -100, when it was around -120, I was convinced that it was in the upward swing toward INCREASE. I threw it to FULL INCREASE well before the normal ground rules required, and the valve went to MAX. Despite the fact that it was in INCREASE, the needle eventually went into the INCREASE position. I don t think we got over a 100. At the end of the burn we were three or four-tenths behind.

ALDRIN Even by leading it as much as I did, I still ended up being a little bit behind. That was pretty small compared to what it could have been.

ARMSTRONG How about the burn itself, Mike?

COLLINS It was just about nominal.

ARMSTRONG Buzz, give the pad value for burn time.

ALDRIN 6 02.

ARMSTRONG Burn time was about 5 57. So it was 5 seconds

ALDRIN Yes. Fairly early in the burn, we could tell that.

COLLINS I remember you were predicting that.

ALDRIN Three or 4 seconds early is what we predicted.

COLLINS Start transient was very small, and steering was extremely quiet and accurate. The chamber pressure, which we had noticed to be a little bit low in the first SPS burn, climbed slowly and actually ended up slightly over 100. I put some specific comments on the voice tape. I thought it was a nominal burn.

7.5 ORBIT PARAMETERS

ARMSTRONG In postburn NOUN 34, we had a 60.9-mile perigee and a 169.9-mile apogee.

7.6 BLOCK DATA UPDATES

ALDRIN   The LOS that we used, in addition to star checks, to tell us if we were in the right position relative to the Moon and the Earth was like the horizon check and is an additional cross check. These calculations turned out to be within a second of the ground-predicted time. When the ground said we were going to lose signal at 75 41 23, it was a second later that signal strength dropped down. It was very comforting. We could see the horizon coming up a good bit before. I guess it was the one for TEI that was a little confusing as to which way we were pointed.

ARMSTRONG   You were the only one confused.

7.8 ADEQUACY OF CONTACT WITH GROUND OPERATIONAL SUPPORT FACILITIES FOR LOI

ALDRIN   Before the burn, I had noticed a difference in the A and B N_2s. I didn t record which one was higher. They were well within what we consider nominal; it stuck in my mind that there was a difference. It wasn t too surprising when the ground called us after the burn and said that they had observed tank B nitrogen had dropped down somewhat during the time of the burn. I think it dropped to 1900.

ARMSTRONG   The values I have are B - 1950 psi and A - 2250 psi postburn. The helium was 1500 psi. Those came up a little bit after the temperature stabilized.

ALDRIN   We ll talk a little more about that. Evidently there was not any particular leak. It might have been a thermal condition that one tank had been exposed to.

ARMSTRONG   The flow through that particular solenoid valve could have been greater than emphasized.

ALDRIN   We started that one on B and then went to A. I don t know if that would be any explanation.

ARMSTRONG   I can t think offhand why that would affect it. The only thing I think about is the size of the orifice through which the gas is passing or the chamber size that, somehow, it was feeding

COLLINS   I don t think we were ever concerned that we had a problem on the B side.

ARMSTRONG   No.

COLLINS   We were glad the ground was looking at it. It seemed to be all right to us.

7.10 ACQUISITION OF MSFN

ARMSTRONG   In the post-LOI, we had a MSFN contact on time and did a P52 option 3 and 2 drift check. Those numbers were reported.

7.15 SPS BURN FOR LOI-2

ARMSTRONG   LOI-2 was a bank A only burn. I assume this was to conserve nitrogen pressure in the B cell. This was a 17-second burn. Residuals were reasonable - 3.3, 0, and 0.1. The DELTA-V_C again was 5.2.

7.17 ORBIT PARAMETERS FOR LOI-2

ARMSTRONG   Postburn NOUN 44 was 54.4 by 66.1.

COLLINS   Did you want to talk about that orbit being targeted 55 by 65 rather than 67?

ARMSTRONG   Yes, I think we made it clear on a number of occasions preflight that we were not in agreement with the change, just prior to flight, to the 55 by 65 orbit. We did not disagree with the intent of what they were trying to achieve; it s just that this did not have the benefit of its effect on a number of other areas of the flight plan. I still feel as though that was somewhat of a mistake. There were some other sides to the discussion that had not been fully reviewed by all parties.

SPEAKER   What about items between the two maneuvers?

ALDRIN   One item that came up was the request to look at the crater Aristarchus to see if we could see any glow or evidence of some observations that had been made by people on the ground. That does bring to mind that as we were coming in on LOI and I could see the edge of the Moon coming back into the daylight, it appeared to me that at one point (which I can t identify) there was one particular area along the horizon that was lit up. I doubt that it was anywhere near Aristarchus. There appeared to be one region that was a little unusual in its lighting. Maybe our films will catch that. We ll just have to try to identify that one when we see the pictures. I don t think that there is any particular connection, but I thought I d mention it because it did strike me as a little unusual.

ARMSTRONG   As long as we re talking about Aristarchus, I d agree with Buzz s observation that the brightest part of the area that was somewhat illuminated might agree with the zero phase point of earthshine. This would mean you re getting a lot of local reflection from earthshine. That certainly ...

ALDRIN   You talking about once we were in lunar orbit?

ARMSTRONG   Yes. I would certainly agree, particularly with the highly illuminated parts of the inside of the crater wall. I think it was also true that the area around Aristarchus, that is in the plains, was also more illuminated.

ALDRIN   It wasn t just the crater, it was the whole general area.

ARMSTRONG   It s not necessarily obvious that this also would happen to agree with the zero phase point of earthshine.

COLLINS   It could. We had nothing to compare it with.

ALDRIN   This was not in sunlight; it was in earthshine. That wouldn t have been zero

ARMSTRONG   Off-hand, it doesn t agree with anything I can think of, and it seemed to extend for quite a distance around that area. Although I called that a fluorescence, it s probably not a very good term. It certainly did not have any colors that I could associate it with. There was just a higher local illumination level over the surface at that point.

ALDRIN   It was a brighter area than anything else we could see in either direction. I don t know if you could compare that with any of the brighter areas we saw in the sunlit portions say on the back side; it didn t look like it was the same thing at all. Not having anything to compare it with in the way of earthshine illumination, we really couldn t tell much.

ARMSTRONG   We could say the effect was there, and it was a very pronounced effect. It s a more obvious effect than looking at the Earth s zodiacal light. It s a more pronounced effect than zodiacal light which is also observed easily with the eye. Our post-LOI-2 P52 option 3 was a good one with an extremely low torquing angle (torqued at 81.05).After this, we prepared the tunnel for LM ingress.

ALDRIN   Let s go back to the first time we went into darkness on the front side, in higher orbit before LOI-2.This was before we got to the region of the landing site. It wasn t illuminated at that point. I guess it s a question of your eyes being light-adapted to the lighter things that you are looking at that are in sunlight.The contrast when going into the terminator was very vivid. There was just nothing to be seen, yet you would wait a short while and then you d pick up earthshine, and you could see quite well. As soon as the sunlit portion of the Moon disappeared from your eyes, you could get dark adapted. Then we could start looking at things like Aristarchus. There was as much earthshine on the dark side of the terminator as there was later on, but your eyes could just not adapt to it, and it was just pitch black. After a short while you would be able to pick up fairly reasonable lighting coming from the Earth. I don t know what you would relate that to, or if you d say that s at all adequate for any landing operations. I doubt that. It certainly did enable you to make observations.

ARMSTRONG   I think that adequately states it.

ALDRIN   We didn t do an extensive amount of observing in earth shine.

ARMSTRONG   I thought it was about 5 to 10 minutes past the terminator before I was really observing things in earthshine very well.

ALDRIN   I think earthshine is four or five times as bright as moonshine on the Earth.

ARMSTRONG   I don t remember making the comparison. It was done on previous flights. Some of the people on previous flights thought it might be conceivable to make landings into earthshine. I don t guess I would be willing to go that far yet. It looked like the amount of detail that you could pick up, at least from orbital altitude, wasn t consistent with what you really need in order to do a descent.

ALDRIN   You might do things like telescope tracking or even sextant tracking. ... characteristic features in the sextant, though we didn t try to do that.

ARMSTRONG   It s difficult to pick out things in earthshine, unless it s a very pronounced feature like Copernicus, Kepler, or some of the bigger craters.You could see those way out ahead and track them continually. For smaller features that are not well identified with large features close by, I don t think you would be able to pick them up. We are ready for the second hatch removal now.

COLLINS   We stored the hatch in the conventional place, that is, in the hatch stowage bag underneath the left-hand couch.That was an easy and convenient place to stow it since they enlarged that bag and it fit very well. It was out of the way.

COLLINS   We stowed the probe, as one of the previous flights suggested, under the right-hand couch with the nose of the probe in the plus-Y direction. It was strapped underneath the foot of the right-hand couch with two straps which were specifically designed to stow it. We just stuffed the drogue in between the LEB and the probe and held it in place with a couple of general-purpose straps. It seemed to work well.

ARMSTRONG I was thinking ahead about our overall LM stowage which was different from our preflight plan with respect to leaving the probe and drogue stowed in the command module overnight.

ALDRIN After LOI-2.

ARMSTRONG Subsequent to this time.

ALDRIN It seemed that all the pluses were in favor of doing that.

ARMSTRONG I agree. I really did not think it was a big thing. We did it to try and save time at the start of the DOI day. We had it removed and it was stowed. That meant that on one night, we had to arrange a sleep configuration with the probe and the drogue stowed in the command module.

SPEAKER Who slept with this?

ALDRIN I did. It was a little cramped under the right seat with the probe and drogue, but I was able to sneak in underneath it. I think I made one exit over the hatch end of the seat. I guess the only thing that leaves you a little bit open to having the probe and the drogue in the command module is if you ve gotten separated from the LM.

7.29 TRANSFER OF EQUIPMENT

ALDRIN In our activation checklist, we have a CSM to LM transfer list. We reviewed this, added a few things, and put some notes on it. I think it would behoove follow-on crews to pay close attention to this type of list, especially if they use this list to record anything that is brought back into the command module from the LM. We brought the purse back in with us. The transfer storage assembly, along with one transfer bag, was used to keep track of everything that was going to be transferred to the LM the next day. We elected to take a few snacks in with us and also added tissues to the transfer list. In thinking about it, I don t believe we had any tissues in the LM.

ARMSTRONG There were, but we couldn t recall where they were.

ALDRIN I still don t recall where they were. We had a couple of towels but we certainly needed the tissues. We found that out the first day we went in translunar; when we pulled the window shades down, the windows were covered with moisture. In order to get any pictures and to test the cameras, we had to bring in some tissues and wipe the windows off. We found considerable use for the two packs of tissues that we took in. I think that is something that ought to be added to the LM stowage.

ARMSTRONG It is probably worth mentioning that, due to various attitude constraints, sun positions, and so forth, you frequently find yourself putting the LM window blinds up and down in lunar orbit. When you put them up, you are going to start to collect moisture on those windows in some attitudes. Invariably, when you take the window shades down, you have partially degraded windows.

ALDRIN It took a long time. You couldn t just wipe it off once; it came right back because the glass had cooled so much.

ARMSTRONG It would clear if it was left exposed to the sunlight for a significant period of time, but we didn t always have that much time before we had to be tracking or looking at the ground or doing something else. Having the tissues or towels there to dry those windows off so that we could use them as windows was important.

ALDRIN   Another item that we added to the transfer list, and we asked for approval from the ground for this, was the monocular. We felt we could use it more in the LM than Mike could in the command module so we took that in with us. We did use it on the surface, looking at and observing certain rocks before and after the EVA. I certainly will recommend that crews have something like that on board the LM, in the way of a magnifying device.

ARMSTRONG   It is useful also before EVA to help plan your EVA routes and objects of interest.

ALDRIN   I might mention that when we went in there the first day, I did go over the circuit-breaker checklist that we were going to do on LOI day and I also went over the complete switch checklist. In essence, we got ourselves 1 day ahead. On LOI day, I went over the circuit breakers but did not go through the complete switch list again. That gave us a little more time to go through the rather brief COMM procedures that we had. I might mention here that the systems test meter in the command module showed that the LM power position was always within limits. It did oscillate rather rapidly between about 0.3 or 0.4 and about 2.2 volts; generally around 1.2. The on and off cycling of the LM loads was much more rapid than I had anticipated.

ARMSTRONG   Every few seconds, the voltage level of the LM bus would change significantly.

7.31 POWER TRANSFER TO LM

ALDRIN   I have logged the times of transfer to LM power, 83 hours even, and transfer back to CSM power, 83:38. The intervening time was spent checking out the COMM. All of this was done on low voltage tap. We checked the OPS pressures both on the first and second days and they were well up there - 5750 and 5800. The REPRESS valve certainly does make a loud bang when you move it to CLOSE. There doesn t seem to be any way to avoid that, especially when you go to CLOSE; it seems you are relieving some pressure. When you go to REPRESS, it is possible that you could avoid it by being very deliberate when you open it. I wasn t able to do it any of the times that I activated it. The COMM seemed to be very loud and clear. I guess that s about it for the LOI day activation.

ARMSTRONG   Just about this same time we had a P22 - our first P22. Comment on that, Mike.

7.33 LANDMARK TRACKING

COLLINS   It went normal. I have on my map the location of the crater on which I marked. I ll give that to the appropriate people. All procedures, the update, the map, the acquisition, everything was nominal.

ALDRIN   I m not sure whether it was this pass or the one before that you were back in the command module and we had a good view of the landing site coming up. I m sure it must have been because we were too busy to be gazing out the window on DOI day. I d recommend that both LM crew members be in the LM on LOI day. Even though you thought you had a good view, I was convinced that I had a much better one than you did.

ARMSTRONG   You probably did.

ALDRIN   ... straight out the window of the approach. I think both crew members probably ought to be in the LM during that time.

7.37 CONSUMABLES ACCOUNTABLE

COLLINS   Because of transposition and docking and P23, we started off behind on RCS and we stayed slightly behind on RCS. The other consumables, oxygen and hydrogen, were within limits. What about LM consumables?

ALDRIN   I guess we went to bed according to the flight plan. How many hours did we have scheduled?

ARMSTRONG   We had a 9-hour rest period scheduled starting at 85 hours.

ALDRIN   I think the reason we were able to be in position to take advantage of the rest period at the beginning of it was because we had already gotten used to the LM operation.

ARMSTRONG   I guess we knew all along that that could be the problem on our timeline, just as it could have been on 10. Any time you get hung up in that DOI day on LM systems, you re not going to make it. We had that same strong inclination to try to be ahead and try to understand the LM as best we could before that time period.

8.0 LUNAR MODULE CHECKOUT THROUGH SEPARATION

<u>8.1 COMMAND MODULE</u>

<u>8.1.1 CSM power transfer</u>

COLLINS   I recall that went without incident. Prior to this time, the CSM had been providing power to the LM heaters, and you could watch the load cycle on the service meter as the heaters cut in and out. They were always within limits. Eleven amps is supposed to be the maximum and I don t think we ever went over two-thirds of that.

ALDRIN   I think it s worthwhile to point out that we didn t jump ahead of the timeline by getting up early. I think we felt confident that the time we had was sufficient.

ARMSTRONG   We had 2 hours before we went in from wakeup. I think it probably worth mentioning that, none of us got as good a night sleep that night as we had the previous night. I m sure it was just that the pressure was beginning to build at this point. We were coming up on DOI day. We got 5 to 6 hours sleep that night. I guess I should have expected that.

<u>8.1.2 Updates</u>

COLLINS   They updated us with the fact that A1 was 500 feet above the landing site. That didn t seem to turn anybody on because the LM charts aren t that accurate.

<u>8.1.3 IMU realign</u>

COLLINS   We did the IMU realign, and it worked okay.

<u>8.1.5 Assist LM VHF A and B checks</u>

COLLINS   I assisted the LM VHF checks, and they worked fine.

<u>8.1.6 Tunnel closeout; probe, drogue, and hatch</u>

COLLINS   Tunnel closeout went normally. The probe, drogue, and hatch worked flawlessly. At this time, I was on this solo book, and the solo book worked well. I went through it and checked things off item by item. The undocking went normal. You may want to say some things about that undocking and station-keeping in regard to who was going to thrust, how it worked out, and what it did to our state vectors.

ALDRIN   We do want to go back and review some LM activities.

ARMSTRONG   Let s go back to the tunnel closeout. As I remember, you were clicking along in good shape there, but we were well ahead over in the LM. We were, in fact, waiting on CMP to get this whole long series of things done. That was completed by clearing the tunnel and getting that ready for us to go. This is a time period when the LM and command module activities are interrelated and dependent on each other. You have to do things in a particular order and be careful that you don t get out of sequence here. That was the first place where we had to sit still and wait.

COLLINS   It did go according to my schedule; it went right along like it should have. There isn t much you can do to hurry that probe and drogue. All that I did in that tunnel I did very slowly and deliberately as per the checklist.

ARMSTRONG   The next thing we did was maneuver to the tracking attitude which you had to do after getting the tunnel all set up to do the P22.

COLLINS   The hooker was I couldn t do that until the tunnel had vented down to a certain pressure level. There is a constraint on 2-jet roll, 4-jet roll, and no-jet roll, depending on the condition of the tunnel. That may have been when you were just sitting there waiting. I had you inhibit roll command until the LM/CM DELTA- P was rated at 3.5. Then I had 2-jet roll started and I was going to start maneuvering to the track attitude. All that timeline went exactly according to the flight plan. If you were ahead, then that was the point at which you had to stop. We did P22 on crater 130 this time, and it was with half-jet authority because you d unstowed your radar antenna. I had to deactivate two thrusters. P22 went just fine.

ARMSTRONG   Then you maneuvered to the AGS calibration attitude.

COLLINS   The AGS calibrate attitude held steady. As far as I know you were leisurely able to get a good AGS cal.

ARMSTRONG   Undocking was one of the things that had to be done very carefully in order to avoid getting some muddled DELTA-V in the state vector from which we could never recover. The procedure that we used was one that was agreed upon within the last week or two before flight. It involved the LM getting up both P47 and the AGS during the undocking time and zeroing the DELTA-V s of the undocking. P47 was one that we chose to zero. As I remember, there was a little residual left in the AGS. Do you remember, Buzz?

ALDRIN   Yes.

ARMSTRONG   We went P47 to zero, and we still had a little left in the AGS. I can t remember whether it was 0.1 or 0.2.

ALDRIN   It was 0.3 or 0.4. It jumped - this was in 470. It just appeared to us that since we had P47 going it was probably the more accurate of the two.

ARMSTRONG   After separating for a distance of 30 to 40 feet - then taking the DELTA-V out in P47 - we asked Mike to choose his own separation distance for watching the gear. He then stopped his relative motion with respect to ours; and the intent was, at that point, both vehicles would have exactly the same state vector that they had prior to undocking.

COLLINS   Any error we had in there might well have been the reason why you might have been long.

ARMSTRONG   Possibly, it may have contributed to that.

COLLINS   I don t know how much error we had in there. I did have to fire lateral thrusters several times and pitch thrusters once or twice. As near as I can tell, those things should have just about compensated for each other.

ARMSTRONG   It was our intention to try and keep the command module from firing any thrusters once he had killed the relative rate. We didn t quite accomplish that.

COLLINS   I didn t have to fire any toward you or away from you, but I had slow drift rates back and forth across you and up and down while you were doing your turnaround maneuver. I had to kill those rates. I don t know how they developed.

ARMSTRONG   The resultant station-keeping was one that was very good. The vehicles were pretty much glued together, 50 to 70 feet apart. How about the inspection?

COLLINS   Inspection consisted of two things, a gear check and a second just looking for any obviously damaged parts or bits of hanging debris. The LM looked normal to me. I had to confirm three of the gears by actually checking the downlocks. I never could get into a position to check the downlock on the fourth gear. I think it was in position initially for downlock inspection but I missed it due to camera activity. Then it rotated around, and I never really could check the fourth downlock. I was relatively confident in saying all four of them were down and locked just by the angle which the gear itself made. All four gears were at the same angle. I took considerable 70-mm as well as 16-mm pictures during this time. If I had spent more time looking out the window and less time fiddling with the cameras, I probably would have had to fire the lateral thrusters and vertical thrusters a little bit less. I called P20 after the separation burn. The SEP burn was within 8 seconds of the flight plan time. I called P20 in that little football we were in, but it was not very accurate. The flight state vector ... make considerable inaccuracy in P20, so the sextant was not able to track the LM. I had been on the solo flight plan book now ever since a GET of 94 hours. This solo book concept, where I had all the information I needed in one book, worked very well. I have no suggestions for any modifications to this book. I used the flight plan as a basis for it and then I inserted more detailed pages during the intervals when the timeline was busy. My original intent in using that approach was that it would be less work for the people who had to make up the book if they could start with something that already existed, like the flight plan. At any time, I could see what was going on inside the LM if I had an inclination to do so. I m not sure it turned out to be any less work. I never was too concerned about what was going on inside the LM, but it did have one great advantage which sort of accidentally fell out. The detailed procedures were done by the McDonnell Douglas people, and the flight plan was done by the flight plan people; and in case after case, the two did not agree. Having them sandwiched in belly to belly immediately pointed out areas where they did not agree. The two groups would then get together and find out why they did not agree. It was a good mechanism for making sure that all counties were heard from. The command module solo activities were exactly in keeping with the flight plans. For that reason, I recommend this particular format.

<u>8.1.21 Rendezvous radar and optics checks</u>

COLLINS   About the only optics checks I got prior to DOI was the fact that I could see the LM through the optics. P20 was not that accurate.

<u>8.1.22 Fuel cell purging</u>

COLLINS   Fuel cell purging was nominal.

<u>8.1.23 Update pads</u>

COLLINS   Update pads were good.

<u>8.1.24 COAS calibration</u>

COLLINS   I did not calibrate the COAS.

COLLINS   I did not track with the COAS. After DOI, I did P20 tracking of the LM. I updated the state vector by using both VHF ranging marks and sextant marks. This is something that was not part of the original flight plan. There was no requirement initially for the command module to track the LM between DOI and PDI. It was something that I added and I m glad I did, because it allowed me to see that the system was working. We had no scheduled checks on it to see that the mark data were incorporated and just generally to prepare for the next day s activities when I would be marking on the LM for real.

COLLINS   This was a fairly busy time in the command module. These procedures were well designed, and I was able to stick with the flight plan. I did get some accepted updates from the sextant marks. They were: 6.1 ft/sec, 7.1 ft/sec, and the third one 3.7 ft/sec. From there on, they were all down below the threshold.

ALDRIN   I ve given you several thoughts on the various things I had to do - where I was going to put things, when we were going to get the LCG s out, when we were going to open them up, and that sort of thing. I think when the time came to do this we didn t have to do a lot of fumbling around. We knew just what to do. There s only one exception to that - our athletic supporters. I had no idea where they were. I thought they might have been in the same compartment with the SCS s and the LCG s. I didn t see them anywhere, and we couldn t see asking the ground where the heck they were. Finally we said to heck with it, and if they weren t there, why we d get along without them. Low and behold, they were inside the LCG s when we opened them up.

ARMSTRONG   I think I commented that is where they were stowed, but when we looked in the LCG s we sure couldn t see them anywhere.

COLLINS   I thought those had been sealed up long ahead of time.

ARMSTRONG   Yes, that s what I remembered, but we sure couldn t prove it to ourselves.

ALDRIN   I don t think there was anything that got me hung up at all in getting a good meal. We knew that we would be going about 6 to 8 hours, at least; so we had a good size breakfast, took care of everything, got up about on schedule, suited up, and stowed things pretty well in the LM.

ALDRIN   Mike had things well under control, and I d been into the LM twice before, so the entry procedure went very rapidly. We were due to go in at 95:50.

ARMSTRONG   We did no complete self-donning. We always used whoever else was available to help with zippers and check wherever they could. We checked each other whenever time allowed.

ALDRIN   I transferred to LM power at 95:54. We did enter the LM right on schedule. We didn t get ahead. I think we had built up enough confidence in the activation procedure by having done this many times in the SIM s. Gene Kranz wanted to run as many of the DOI and PDI SIM s as we could, starting right from activation, and I think it was a good thing that we did. Leaving the simulator run, we found that we had plenty of time to go out and get a cup of coffee or make a phone call and get back in again. Having gone over this many times, we had the confidence to go ahead and not try to jump ahead. I think that things worked out quite well. We were gradually, comfortably getting 15 to 20 minutes ahead. I d liked to have delayed going to the high-voltage taps and activation. Page 19 says to go ahead and do that and get the bus voltages below 27, but they weren t. I don t recall the exact time in the checkout when they did begin to approach 27. I think it was during the circuit breaker activation when we put everything on the line. It was about that time that the voltage started going down. Here is an example of how our time schedule went: caution and warning checkout was to start at 96:41, and at the latter part of that is a step called primary evaporator flow 1, open. I logged 96:05, as the time we opened that. At that point, we were 30 minutes ahead. In the circuit breaker activation, the only funny that I observed was in putting the LGC DSKY circuit breaker in. We had a program alarm 520 on the DSKY; 520 is radar erupt, not expected at this time, and I can t explain that. We reset it. We didn t have any radar on. We ll just have to see what the people say about that. I think that Neil came in just about on schedule. I was able to accomplish three or four headings that we were going to be doing together. I had to wait until he got in before doing the suit pan water separator check. He had to be hooked up at that time. It appeared as though it wouldn t be wise to get that one out of sequence. I did get the glycol pump check at 97:05. I recorded that Neil was in. By the time he came in, I was to the point where I was ready to go back in and put my suit on. That got me something in the vicinity of 15 to 20 minutes ahead. I knew we pretty well had it made at that point. We did E-memory dump, and you did some work with the DSKY and the alignment checks.

ARMSTRONG   The E-memory dump was repeated for some reason or other. I think we lost S-band.

ALDRIN   I can t recall if it was an attitude problem, but we did do that again.

ARMSTRONG   For some reason, we lost the high bit rate during this time period. The VHF checkout was good. Both VHF A and B between the two vehicles were good. The time and TEPHEM initiations were without problem, and we did the docked IMU coarse align. The advantage of being slightly ahead showed itself in that MSFN was able to compute the torquing angles before we lost signal with them, before we went on the backside. They gave us the torquing angles, and we torqued the platform at 97:14, about an hour before we were scheduled to do the initial torquing. This gave us better drift checks, which was a help in analyzing the LM platform. We had never done that in the SIM s. Later on I was a little confused in my own mind as to what cages that might result in and whether we would have the subsequent torquings about an hour a half later. At this time, I had nothing really further to do until Buzz returned with his suit on. When he came back, we only had to wait on Mike to get the tunnel closed up before we could continue with things like the pressure integrity checks and regulator checks.

ALDRIN   It seemed to me we spent a good bit of time holding at just about that point.

ARMSTRONG   We were a little ahead, and it turned out that there was very few things that we could do or wanted to do at that point.

8.2.5 ECS

ALDRIN   The glycol pump sure made a lot of noise.

8.2.13 Ascent batteries

ALDRIN   You kind of hate to bring the ascent batteries on the line. You ve got a system going and then turn off all the descent batteries just to prove that the ascent batteries are working. You have no backup if you turned off all the batteries; at that point, everything would go dark. Maybe that wasn t the only way you could go about checking to see that the ascent batteries worked. But, that worked out all right.

8.2.15 ARS/PGA pressure

ALDRIN   The pressure integrity check held with the suit loop decreasing maybe 0.1 or something like that. I think there is a little lag in there when you first close the regulators. The tolerance is 0.3. It wasn t anywhere near that. There wasn t any significant change going to the secondary canister. The regulator check is a fairly involved setup of valve switching. I m sure all of these things are nice to do, but unless you have an extremely intimate knowledge of exactly what you re doing, you can run into some problem there. The fact that you re doing this one step right after another puts you in a non-nominal situation. I would much prefer that this sort of a check be done on the Earth side where you have COMM, because you re dumping the cabin pressure down and you re using a REPRESS valve. I think the ground would agree with that, too. If in other flights it could be worked into the earth-side pass, I think it would be beneficial.

ARMSTRONG   I agree with that, although I think the pressure integrity check is relatively straightforward. But these two are coupled together. It s tied to the tracking and to the tunnel closeout.

8.2.16 AGS activation, self-test, calibration, and alignment

ALDRIN   We had already had the platform up and it had been aligned to the command module s platform, so I went through the AGS initialization update. I knew that we didn t have a state vector, so there wasn t any point in putting the state vector in. I was smart enough at that point to recognize this and I knew that the state vector was coming up later. But I thought, Well, there s nothing to stop me from aligning the AGS platform to the PGNS platform, so I did this and immediately looked at the AGS ball and it was way out in left field. It didn t agree to the PGNS ball at all, and it took me about 5 minutes or so to try and figure out why this was. I finally realized that the reason for it was that the PGNS didn t have a REFSMMAT, and its computer didn t know where its platform was. Even though the platform was in the right spot, it didn t have any reference system so it couldn t tell the AGS what its platform ought to be. The AGS platform, in terms of the command module, is in the forward plane. The PGNS didn t know this. It just came up with some garbage. Well, this caused a little bit of concern because we were quite anxious to have the AGS with us for the whole flight. We were beginning to wonder whether we would or not. Let s see, there was one funny thing that I don t think we ve mentioned. It was pretty minor. One of the strokes on the DEDA was not illuminated. Each character is made up of all these different strokes. The one missing was in the middle character, and it would leave you in a position where you couldn t tell whether it was a three or a nine. I didn t realize at the time that there was any room for confusion. Later, in looking at some numbers, you could not really tell whether in fact that was a three or a nine.

ARMSTRONG   Yes. You just need that one stroke to close it, and it becomes a nine.

COLLINS   I got the bottom one.

ALDRIN   With this particular one missing, there was some doubt as to exactly what you had.

ARMSTRONG   That s true of any digit on any of those electrical switch displays.

COLLINS   Remember, we had one of those in the EMS.

ARMSTRONG   Yes, that s right. Fortunately, the simulators usually got some out and you got used to putting up with that. But, it s a problem that really could get to you some time if you misinterpret that number.

ALDRIN   We missed putting the AGS time in there. We missed by 15 centiseconds hitting it right on, which I thought was very close. We did even better than that when we updated at 120 hours.

8.2.17 S-band antenna

ALDRIN   The S-band antenna seemed to work very well at this stage. It didn t make quite as much noise as I had anticipated.

ARMSTRONG   However, it was noticeable.

8.2.18 ORDEAL

ARMSTRONG   As we set up the ORDEAL, we got back to our favorite argument. That is, what is right to set in the ORDEAL the AGS or the PGNS, when you re at nonzero yaw? I guess we believed that it was the AGS that was right. We set it in and as it turned out, it was right and the PGNS was wrong. By about 40 degrees or something like that.

ALDRIN   The PGNS was wrong by 40 degrees.

ARMSTRONG   That s an interesting one, because you can get either answer depending on who you ask; I still think that today. We at least proved to ourselves that the AGS was the correct one.

ARMSTRONG   Landing gear went down very nicely. No problem with the landing gear and there was no question about that one.

8.2.19 Deployment of landing gear

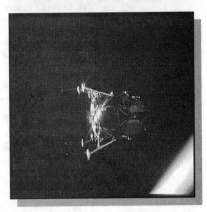

ALDRIN   We were expecting two distinct sounds, but really they weren t identical sounds. You could hear the PYRO s fire, and just a short time after, there was not as much sound as there was a vibration transmitted up that indicated something had locked down. Of course, we had no way of knowing how many of them had done that. However, when we did fire, we opened up logic power A when we fired them, and then we closed logic power A and fired again, and at this time we heard a click just like a relay going, but no PYRO s fired.

ALDRIN   Now, how about the gimbal trim.

ARMSTRONG   We did not drive the gimbal. Some question arose while we waited for confirmation from the ground, but they had proper gimbal positions, and we did not have to drive.

ALDRIN   I recall no problems there. The parker valves in the talkbacks gave us some rather funny responses. Gene s comments indicated that when you activate one of the quad pairs or main shutoff valve to a particular position, it didn t go to that new position until you released it. Through most of our training in the simulator, you d move that valve as soon as you d get it to the spring loaded position of open and close, it would change, and it would stay changed when you would release it back to the center. If it didn t work that way, when you moved it, it didn t go to its new position until you released it. So we changed the simulator. We found something even further than that. The ascent feed l s were open, and the 2 s were closed. All of them were barber-poled as we expected. After pressurization, the procedure was to go through and cycle each valve to its present position, where it should be. So, I went to the ascent feed l s and went open, and nothing changed; it stayed barber-pole. As soon as I went to number 2, the closed position which would put them barber-pole, they both went direct. They went to the opposite position that you would not expect. When I released it, they went back to barber-pole again. I think the same thing happened to the shutoff valves. When you d move it to the closed position, where it should go barber-pole, it would go gray. Then, as you released, it would go to its present position. You can t tell the position of the valve until you release it. As a matter of fact, it ll give you the opposite indication in some cases. We had good helium pressure and read that out to MSFN. We went through the RCS checkout. We had one quad, upper right-hand one, that stuck two different times in the red indication. When going through the cold fire, we were getting all different stories from the ground as to whether these talkbacks would go red. The final one that I got was, No, the latest story is they won t go red on you. Well, they all went red. First four of them, then all of them went red. It s a very light-colored red, I might add. It didn t look much like the simulator. It really stands out much more than the simulator. We got the numbers we ran on the DSKY when we went to the soft stops. For the most part, they agreed precisely. There were a couple of them that missed by one last digit, but we were told that that was not significant.

ALDRIN   Everything went just as expected. I ve got the numbers written down here; they re all within limits.

ARMSTRONG   DPS preparation and checkout went as expected.

ALDRIN   The AGS CAL attitude angles are written down in my log. Mike maneuvered to the angle, and we re steady as a rock for a good long time period; more than adequate time period to perform the check.

ARMSTRONG   I d always wondered if there was anything that you could do during the AGS CAL 5-minute period that would maybe give a little jolt, go back to the AGS, and give you an erroneous reading so you wouldn t pass. In any case, we just avoided that problem by not doing anything except the AGS CAL during the AGS CAL. We didn t pressurize the DPS, or put down the landing gear, or run the rendezvous radar or any of those things which might put a little oscillation into the spacecraft and trigger an accelerometer or something of that sort that might cause a problem. We just let it run all by itself.

ALDRIN   This pressurization sounds like a big thing, but really it took about 2 minutes to do.

ARMSTRONG   Yes.

ALDRIN   And we went through the final circuit breaker verification. Cards worked quite well. We d lose maybe a little bit of time by having to pass them back and forth. I don t think that s too significant.

8.2.28 Undocking

ARMSTRONG   Undocking was very smooth. We had a very good visual. We could always tell where the command module was by looking out the window. We commented on our concern about the manner in which the undocking was controlled. I think there s still room for improvement on that procedure. One that was discussed before flight was: extend the probe, and then release the capture latches - essentially have no velocity between the vehicles. Then the command module really wouldn t move at all at the time we clear away and wouldn t compromise the state vector in any way. We thought that might be a very good way to do things but we just didn t feel that there was enough time before launch to look into the secondary effects you might get out of doing something like that, so we chose to go with the way undockings had been performed previously. That may be something future flights might want to look into with more care than we were able to.

ALDRIN   Putting the helmet and gloves on and off didn t seem really to be much of a bother. We put them on for the integrity check, took them back off again, put them back on for undocking, and took them off. The little piece of Velcro on the feet port worked quite well, just slapping it down on the ascent-engine cover. I put my gloves over by the right-hand controller. You could put them in the helmet just as well.

ALDRIN   The verification of about 8 to 10 AGS addresses I was able to get done before undocking. There is a bad amount of data that the ground reads up to you in that time period - the DOI, PDI, PDI plus 12 pads, and various loads that are coming up. You have to devote one man just to copying all those things down. It seemed like it took forever to get them all done. Even after we got those, we still had some more coming up after DOI; the surface pad had to come up.

8.2.33 Formation flying

ARMSTRONG   Formation flying was considerably less difficult than our simulation would lead us to believe. We were able to maintain position with respect to the other vehicle. It was less trouble than in simulations and used less fuel. At separation, we thought we had relative velocity nulled to less than 0.1 ft/sec in all axes. This was based on the size of the translational inputs required to maintain a constant position over past 10 or 15 minutes before separation.

ALDRIN   We did add 20 degrees to our pitch attitude after undocking, so that we d get better high gain during the yaw maneuver. That, I think, is peculiar with the particular landing site, but we were able to get high gain lock-on. As a matter of fact, I could have gotten it before we made the pitch maneuver, but it didn t look like there was too much point in doing that. As soon as we finished the pitch maneuver, we had high gain lock-on and had it throughout the yaw maneuver. I was going to take some pictures with the 16-mm camera mounted on the bracket, but it looked like it was canted off to the side. No comment at all on using the AGS for this versus the PGNS. We made a change from MAX deadband to MIN deadband. This to me is an open area.

ARMSTRONG   We were in AGS, ATTITUDE HOLD, MIN deadband, and PULSE in the axis that we were maneuvering in. The separation attitude was not the attitude we had expected to be in as a result of some changes to the ephemeris at this point. In other words, Mike was separating on the local vertical, but that was not at the same inertial pitch angle that we expected to be at. It was off by about 10 degrees as I recall.

COLLINS   No, SEP occurred within about 8 seconds of the planned time.

ARMSTRONG   You did separate on the local vertical? The pitch attitude that we were at was about 10 degrees different.

COLLINS   It was a 7-degree-different attitude. It was pitch 007 instead of pitch 014.

ARMSTRONG   I was holding in the attitude that was on our timeline, and sure enough, it didn t look like you were in the right attitude. Some changes occurred after launch that we didn t properly appreciate. In any case, 285 is what we expected to be. That wasn t the right number. That was important, because it was the thing that made the COAS point at you and check lateral translations, comparing the formation flying during separation. Immediately after this, we did the landing radar test, right after your separation. That went well, as I remember, every time.

ALDRIN   Yes, they were right on.

ARMSTRONG   After that, we did our first alignment in the LM, fine align, P52, option 3. We did that on the flight plan stars, Acrux and Antares. The torquing angles were about 0.3 degree.

ALDRIN   Yes, and they sure didn t instill a lot of confidence. This was the last alignment we were going to have, and we changed what we had by 0.3 degree. I guess that s to be expected, but I was sure hoping to have smaller ones than that. This indicated the kind of drift we had from the last alignment from the command module, and it was my understanding that these alignments were quite good - better than these torquing angles would indicate.

ARMSTRONG   We re interested in finding out what the drifts were there; whether that was just an inability to calculate any biases and put them into the computer so that you could improve the platform up to what we normally would expect.

ALDRIN   We had a manual lock-on with the radar before we did this.

ARMSTRONG   Yes.

ALDRIN   We had P20 standing by, but we didn t use it at all.

ARMSTRONG   We had a manual lock-on and our radar needles and COAS agreed very well. This was your first chance to look at the transponder and all that stuff in operation.

COLLINS   Yes. I recall I gave you some ranges. I didn t write them down.

ALDRIN   It agreed with our values very closely.

COLLINS   And they agreed with my state vector. I just wrote down one value which was fairly close to yours. When I had you at 0.72 miles on VHF ranging, my state vectors indicated 0.62.

ALDRIN   That was close. Did we get that alignment finished? It seemed to me it took a little longer.

ARMSTRONG   Yes, we took five marks on each star, and it did take us quite a while.

ALDRIN   Yes, I would like to emphasize to subsequent crews to allow lots of time in their timelines when they re doing the alignments. We made a practice early in training of leaving the TTCA switches disabled as much as possible, and the direct coils 4-jet active. I m not sure everyone understands why you do that. It s a good sound thing, I think, to keep as many hand controllers out of the loop as you can. It makes troubleshooting far easier and it minimizes the number of problems you can get into.

ARMSTRONG   It s just a basic difference in philosophy. Most of our Directorate takes the viewpoint that you leave everything on, and essentially everything is hot all the time. We took just the opposite approach; namely, we turned all the things off that we didn t think were contributing, particularly in the control system. We isolated that many more possible failures causing us difficulties en-route.

COLLINS   We did the very same thing in the command module in that we used hand controller number 1 as a spare. We never powered it up and left it alone.

ARMSTRONG   A lot of people didn t understand about disabling this and disabling this switch. It was really just a matter of preventing failures from getting to us in critical times.

9.1 COMMAND
MODULE

9.0 DOI THROUGH TOUCHDOWN

9.1.1 LM DOI
burn

COLLINS   I didn t have any monitoring to do other than just confirming that they did it on time and that it was normal which it was.

9.1.2 AUTO
maneuver to

sextant tracking

COLLINS   I did that, and lo and behold, the LM was in the sextant. This is a good exercise to do between DOI and PDI. It gives you an opportunity to make some sextant marks, make VHF marks, and then to see these marks incorporated into the state vector. It s a good end-to-end test of the whole system.

9.1.3 MSFN
acquisition

COLLINS   No problem

9.1.4 Optics
track - ease of
tracking LM

COLLINS   The LM was easy to track. AUTO optics worked well, and the optics drive was extremely smooth. When using resolve and in low speed, it was easy to take accurate marks on the LM. The LM, of course, got smaller and smaller, and out at about 100 miles, it became quite difficult to see the LM through the sextant. The LM would appear to be just a tiny little dot of light which was easily confused with many other little dots of light on the optics. One trick that you can use is switch from AUTO to manual and slew the optics up and down and left and right. All the other little dots that are associated with the background of the surface will remain fixed, and the LM will then move across them; and you can pick out which little dot is the LM by the fact that the LM has motion relative to the background. This technique works for another few miles, but I don t know how long I could have kept the LM in sight. I lost it prior to PDI when I switched from P20 to P00. My procedures called for me to do this, and in the simulator it worked quite well; however, in the real world at the instant I called P00, I went VERB 37, ENTER 00 ENTER. That stopped the P20 rate drive, and despite the fact that I was prepared for it and was looking through the sextant, the instant the computer went to P00 and the rate drive stopped, the LM just disappeared from view. It took off for parts unknown at a great rate of speed and disappeared to the 6 o clock position in the sextant and at an extremely rapid rate. It was impossible to bring it back, and I never saw the LM again throughout the descent or on the surface or during the ascent until after insertion.

9.1.6 Voice
conference relay

COLLINS   We didn t use the relay mode at all, although I had a little sticker made for panel 10 which showed the position of each switch. I think that s probably a good scheme because if you want the relay mode, you want it in a hurry; and you don t want to pull a checklist out, so I d recommend that.

9.1.7 CSM backup pad

COLLINS   Nothing to say about that. I, of course, used P76 to inform my computer that the LM had made the burn.

9.1.8 Monitoring LM phasing

COLLINS   We didn t have a phasing burn.

9.1.9 Sextant marks

COLLINS   I ve covered those.

9.1.10 SPS setup

COLLINS   For all burns, I was to go into P40 or P41 as appropriate. I then went to the point of turning on the gimbal motors and stopped short. I never turned on any gimbal motors, but I did feel that I could light the motor within probably a matter of a few seconds after being informed that the LM had not made the burn.

9.1.11 Monitoring and confirming LM DOI

COLLINS   After I went to P00, I lost the LM. This was a couple of minutes before PDI ignition. I just went ahead open loop. I followed my attitude timeline in hopes that I could see the LM again. I did my pitchdown maneuver that the flight plan called for. I did that as a VERB 49 maneuver, and it worked fine in that I had a good unobstructed view of the lunar surface, including the landing area and all that; but again I never saw the LM, so for future flights, I don t really know what to recommend. At the beginning of PDI on this flight, the LM was 120 miles in front of the command module, and touchdown was like 200 miles behind the command module; so the geometry is changing extremely rapidly, and there is no automatic program in the computer for helping you track. You had to abandon P20 prior to PDI, and I don t really have any helpful suggestions. The only thing I can say is to be aware of the fact that when P20 is terminated, the LM is going to depart very abruptly from the sextant field of view.

9.1.12 LM tracking

COLLINS   However, if you are all poised and are in resolved medium speed and switch from AUTO to manual at the instant the computer switches from P20 to P00, there is a faint chance that you might be able to track the LM during PDI manually and during the descent. I tried to do this, not because there was any real requirement to do so, but just because I felt that it would be a good initial condition for an abort if I were able to see the LM in the sextant.

9.1.13 Lunar surface flag

COLLINS   After the LM landed, I set the surface flag. There was no evidence ever of any flash of specular light or anything like that off the LM. The LM, at distances of 100 miles or so, is just another little light, little lunar bug that was indistinguishable on the background surface. The surface is pockmarked with little irregularities - light spots, dark spots - and with P20 driving so as to hold that background surface relatively constant and at those distances, you just can t pick the LM out.

9.2 LUNAR MODULE

9.2.1 Preparation for DOI

ALDRIN   It was 40 minutes before DOI that we were scheduled to begin the P52 and we were about 2 minutes behind when we completed looking at the radar and VHF ranging and designated the radar down so that we could do the P52.

ARMSTRONG   I don t think we had any difficulties with the DOI prep.

9.2.2 DPS/DOI burn

ARMSTRONG   At DOI ignition, which was our first DPS maneuver, I could not hear the engine ignite. I could not feel it ignite, and the only way that I was sure that it had ignited was by looking at chamber pressure and accelerometer. Very low acceleration.

COLLINS   I would think under zero g, it would throw you against your straps, one way or the other.

ARMSTRONG   We re pulled down into the floor with the restraint, and the difference between that and the 10-percent throttle acceleration was not detectable to me, However, at 15 seconds, when we went to 40 percent, it definitely was detectable.

ALDRIN   On the restraints, I found that instead of being pulled straight down, the general tendency was to be pulled forward and outboard. So much so that this might have been a suit problem, as my right foot around the instep was taking a good bit of this load, being pulled down to the floor. It did feel as though the suit was a little tight. Prior to power descent, the problem was obscured from my mind, but it was aggravated somewhat by the restraint pulling down and forward.

ARMSTRONG   I guess I noticed that last. I had expected a good bit of lateral shifting due to reports of previous flights.

ALDRIN   I was able to lean over and make entries on the data card without pulling it down; but as you can see, when you do make entries on them, you make them sideways.

ARMSTRONG   The cut-off was a guided cut-off. What about the residuals?

ALDRIN   We burned both X and Z, and I m sure they weren t in excess of .4.

ARMSTRONG   It was less than 1 ft/sec, but I don t recall the tenths.

9.2.6 Trimming residuals

ARMSTRONG   It s probably worth noting that the flight plan at this point does not adequately reflect the time requirements of the flight. I think the DOI rule in the flight plan says, Trim V_X residuals.

ALDRIN   So does your checklist.

ARMSTRONG   That isn t right. This was a result of that orbital change that was put in late, and paperwork and so on just couldn t keep up with those last-minute changes. But, again, it shows that last-minute changes are always dangerous. You could follow the flight plan here and possibly foul up the procedure. Do you recall the VERB 82 values? 9.5 was perilune, I think.

ALDRIN   Preburn for NOUN 42 was 57.2 and 8.5. We had 57.2 and 9.1 after the maneuver.

ARMSTRONG   I guess we can t account for that.

ALDRIN   No. The NOUN 86 that we got out of the thrust program also differed from what the ground gave us in the pad, primarily, in the Z-component that s loaded into the AGS; that pad value is 9.0, and the computer came up with 9.5. The coordinate frame that you load them in is frozen inertially, and if there are any discrepancies in the freezing of this, you will get a slightly different burn direction required out of the two guidance systems. I think that explains the larger AGS residual in the Z-direction of minus 0.7. I think we would have to have the guidance people verify that the difference in NOUN 86 produced that error in that direction.

9.2.9 Radar tracking

ARMSTRONG   We had a good manual radar acquisition, and data from the radar agreed well with the VHF ranging information.

ALDRIN   Again, we had P20 in the background, but we didn t use it. This was a manual lock-on.

ARMSTRONG   The radar was depowered to cool during the DOI to PDI phase.

ARMSTRONG   The platform drift check, a P52, was done against the Sun. This procedure seemed to work as we had planned; however, the variation in the data was somewhat larger than I would ve guessed. Do you have those numbers?

ALDRIN   Yes. The technique that we used was to compare what the computer thought the little gimbal or the inner angle was and to point the rear detent at the Sun. We d compare that with what the actual middle gimbal was. Now we did this in PGNS pulse. The way that we found to work out best was for Neil to tell me when, in the background, we d have the AUTO maneuver display 50 18 in P52. We d call up on top of that VERB 6 NOUN 20 or 22. And I d have NOUN 20 up. As soon as Neil would say MARK , I d hit ENTER, record NOUN 20. Now the desire is to find out exactly what the computed value is in a close time period. So what I would do is hit the ENTER on the NOUN 20, visually recall what those numbers were, not write them down, but hit KEY RELEASE, which put me back to the 50 18 display. A PROCEED would recompute the numbers or maneuver. As soon as I would do that, those numbers would be frozen and the desired gimbal angles would be loaded in NOUN 22. Then it was just a question of my calling them up, and they should not change the time I hit ENTER to record the gimbal angle that we had until it was recomputed as a desired one that did not exceed 3 seconds. Of course, we had pretty low rates. So I think that the comparison didn t suffer any from a lack of proper procedure, we did find that the numbers were a little larger than we thought they would be. We had it worked out with the ground how we arranged the signs on the differences, so we d subtract NOUN 22 from NOUN 20. The first one was 0.19; second one, 0.16; and the third one, 0.11. The GO/NO-GO value was 0.25. So we re a little closer to this than we had hoped to be.

ARMSTRONG   The simulator is able to reproduce correctly the control modes that are required to fly it. It s an unusual control mode wherein you fly to in pitch and fly from in yaw. While flying AOT, you depend on the other crewmember to assure you that the roll gimbal angle is staying at a reasonable value. The simulator was never able to simulate accurately what you would see through the Sun. We especially set up the AOT on the G&C roof (MSC) to look at the actual view. In addition, on the way to the Moon, we looked at the Sun with the telescope; looked through the CSM telescope with the Sun filter on to get used to what the filtered view of the Sun would look like in the optics. It s somewhat different in the telescope than in the AOT in color and general appearance. I can t account for that, but it is different.

I thought the numbers ought to be both closer to zero if we didn t have any platform drift, or closer together in either case. But we had quite a spread, so I m not sure that the check in general is really as good yet as it should be. In other words, our variation was 0.08 degree between our various measurements. The limit on the GO/NO GO is 0.25. So, we were essentially using up a third of our margin just in variation between our marks. That is not really a good enough procedure for this important check of the platform. This procedure, being a GO/NO-GO for the PDI needs additional work prior to the next flight.

There are some alternative methods of understanding platform drift, which we just did not have time to implement. Perhaps the next flights will be able to look at some of these alternatives and decide on an even better method than the Sun check.

ALDRIN   We turned the propellant quantity on before DOI and I believe the quantity light came on at that point, which was expected as a possibility. Just recycling the switch off and back on again would extinguish the light. The values that we saw in fuel were about 94 and 95, which is what we generally saw in the simulator. The oxidizer value was somewhat lower than that. The simulator values were 95 and 95. I don t believe that there was sufficient time during DOI for these to settle down completely. They did approach the maximum numbers with a reading of approximately 94. Anyway, they weren t dancing around the way we might have been led to expect them to do.

ARMSTRONG   The pre-PDI attitude prevented good S-band high gain contact. We had continual communications difficulty in this area until we finally yawed the spacecraft right between 10 and 15 degrees to give the high gain antenna more margin. This seemed to enable a satisfactory high bit-rate condition, but it did degrade our ability to observe the surface through the LPD and make downrange and cross-range position checks. I don t think that our altitude checks were significantly degraded.

ALDRIN   I can t explain why we had some dropouts there. The angles, 220 in pitch and yaw 30, are not ones that would lead you to believe they would give you trouble as far as interference from the LM structure. It seemed to me that the initial lock-on was not bad. There is a certain rain dance you had to go through each time you d come around to acquire lock-on. Each time you d have LOS, we d usually be on the OMNI s. Of course, there s a choice of forward or aft, then you d want to switch to SLEW and slew in the proper values for the steerable. Before LOS on the other side, the ground would like you to not break lock in the slew mode, because in some cases the antenna would then drive into the stops, so approaching LOS, you d switch to maybe the aft OMNI and then you d slew in some new numbers. We d make use of pitch 90 and yaw zero, to keep the antenna away from the stops. Once you drive it to those values, then you d have to set in new numbers.

Coming around on the other side, you d maybe switch from aft to forward to pick up the ground. Once you picked them up, you d switch over to SLEW and you might have the right values down there or you might have to tweak them up. In any event the initial contact would be made on one antenna; and then after you establish contact, you d have to take the chance of breaking it, to switch over to the high gain. Occasionally, we got the jump on them a little bit because the ground was talking to the command module. We saw that we had signal strength so I d go ahead and try to lock on the S-band. It is a rather involved process that you have to go through. I didn t find that, if you left the antenna without an auto lock-on signal, it would have a tendency to drive to the stops. At least from the indications, it didn t seem to be moving so rapidly that you couldn t, within several seconds if you knew what you were doing, stop it from where it was going and prevent it from hitting the stops.

We had two methods of computing altitude: one based on relative motion from the CSM and the other based on angular rate track of objects observed on the ground. We superimposed the two of them on one graph and rearranged the graph a little bit with some rather last-minute data shuffling to give us something that the two of us could work on at the same time and to give indication of what the altitude and its time history appeared to be. With the communications difficulties that we were experiencing in trying to verify that we had a good lock-on at this point, I had the opportunity to get only about two or three range-rate marks. They appeared to give us a perilune altitude of very close to 50,000 feet, as far as I could interpolate them on the chart.

Those measurements give you altitude below the command module, essentially. And, of course, there are some modifications of the command module orbit, from the nominal preflight orbit that you expect. The numbers either have to be updated or you have to accept the error.

ARMSTRONG   The measurements against the ground course were indicative of altitude directly above the ground.

ALDRIN   The main purpose of the radar here was to confirm that we were in the same ballpark, the same kind of an orbit, And I think once you accomplish this several times, then it s adequate to go on with the truer altitude measuring device, which is from the ground.

ARMSTRONG   The ground measurements were very consistent. If they made a horizontal line, it would indicate that you were going to hit a particular perilune, in this case, 50,000 feet (in the middle of the chart). They didn t say that. They were very consistent, but they came down a slope, which said finally that our perilune was going to be 51,000 feet. It steadied out at about 54,000 feet here at the bottom and our last point was 51,000 feet. This indicated that either the ground was sloping; and, in fact, it was about 10,000 feet lower than the landing site where we started (which is not consistent with the A-1 measurement that we made), or that the line of apsides was shifted a little bit. So actually perilune was coming a little bit before PDI. So we were actually reaching perilune a little bit before PDI, which would tend to slope the curve that way. This was all very encouraging that we were, in fact, going to hit the guidance box so far as altitude was concerned from both measurements (the radar measurements and the ground measurements). But I was quite encouraged that these measurements, made with the stopwatch, were consistent, in fact.

ALDRIN   When you re able to smooth the numbers and plot a reasonable number of them, your accuracy increases considerably. I think the preflight estimates were something on the order of a 6000-foot capability, and I think we demonstrated a much better capability than that.

9.2.17 PDI burn

ARMSTRONG   Our downrange position appeared to be good at the minus 3 and minus 1 minute point. I did not accurately catch the ignition point because I was watching the engine performance. But it appeared to be reasonable, certainly in the right ballpark. Our cross-range position was difficult to tell accurately because of the skewed yaw attitude that we were obliged to maintain for COMM. However, the downrange position marks after ignition indicated that we were long. Each one that was made indicated that we were 2 or 3 seconds long in range. The fact that throttle down essentially came on time, rather than being delayed, indicated that the computer was a little bit confused at what our downrange position was, had it known where it was, it would have throttled down later, based on engine performance, so that we would still hit the right place. Then, it would be late throttling down so that it would brake toward a higher throttle level prior to the pitchover.

9.2.24 Final
approach and
landing

ARMSTRONG   Landmark visibility was very good. We had no difficulty determining our position throughout all the face-down phase of power descent. Correlating with known positions, based on the Apollo 10 pictures, was very easy and very useful.

ALDRIN   As I recall, there was a certain amount of manual tracking being done at this time with the S-band antenna. During the initial parts of power descent, the AUTO track did not appear to maintain the highest signal strength. It dropped down to around 3.7 and the ground wanted reacquisition so I tweaked it up manually. I got the impression that it was not completely impossible to conduct a manual track throughout powered descent. You d not be able to do very much else besides that. I think it would be possible to do, if you had sets of predetermined values that you could set in.

We did have S-band pitch and yaw angles immediately following the yaw maneuver, and those that were acquired at about 3000 feet. After the yaw, the S-band appeared to have a little bit better communications. It was just about at the yaw-around maneuver (trajectory monitoring from the DSKY up to that point agreed very closely especially in H-dot and VI with the values we had on the charts). It was almost immediately after yaw around that the altitude light went out, indicating that we had our landing radar acquisition and lock-on.

ARMSTRONG   The delta altitude was - 2600 or 2700, I believe, is the number that I remember. I think it was plus 2600 or 2700. The yaw around was slow. We had inadvertently left the rate switch in 5 rather than 25, and I was yawing at only a couple of degrees per second as opposed to the 5 to 7 that we had planned. The computer would not hold this rate of say, I to 2 deg/sec. It was jumping up to 3 degrees and back, actually changing the sign and stopping the roll rate. It was then that I clearly realized that we weren t rolling as fast as was necessary and I noted that we were on the wrong scale switch, so I went to 25 and put in a 5-deg/sec command and it went right around. However, this delayed it somewhat and consequently we were in a slightly lower altitude at the completion of the yaw around than we had expected to be, so we were probably down to about 39,000 or 40,000 feet at the time when we had radar lockup, as opposed to about 41,500 that we expected to be.

ALDRIN   There are no discrepancies noted in any of the systems that were checked throughout the first 4 minutes. The RCS was suprisingly high in its quantity indications. The supercritical did tend to rise a little bit after ignition and then it started back down again. I don t recall the maximum value that it reached. I guess the first indications that we had of anything going wrong was probably around 5 minutes, when we first started getting program alarm activities.

ARMSTRONG   We probably ought to say that we did have one program alarm prior to this; sometime prior to ignition, that had the radar in the wrong spot. In any case, as I remember, we had a 500 series alarm that said that the radar was out of position, which I don t have any way of accounting for. Certainly the switches were in the right positions. They hadn t been changed since prelaunch. But we did, in fact, go to the descent position on the antenna and leave it there for a half a minute or so, and then go back to AUTO and that cleared the alarm. After 5 minutes into descent, we started getting this series of program alarms; generally of the series that indicated that the computer was being overloaded. Normally, in this time period, that is, from P64 onward, we d be evaluating the landing site and checking our position and starting LPD activity. However, the concern here was not with the landing area we were going into, but rather whether we could continue at all. Consequently, our attention was directed toward clearing the program alarms, keeping the machine flying, and assuring ourselves that control was adequate to continue without requiring an abort. Most of the attention was directed inside the cockpit during this time period and in my view this would account for our inability to study the landing site and final landing location during final descent. It wasn t until we got below 2000 feet that we were actually able to look out and view the landing area.

ALDRIN   Let me say something here that answers the question that we had before about the AGS residuals on DOI. They were 0.1 before nulling and we nulled them to zero. X was minus 0.1, Y minus 0.4, Z minus 0.1, and we nulled X and Z to zero. Looking at the transcripts, we did have considerable loss of lock approaching PDI. And we did have to reacquire manually several times. It looked like we had some oscillations in the yaw angle on the antenna. The alarm that we had was 500 and we went to descent 1 and proceeded in the computer and then went back to AUTO again on the landing radar switch. This was prior to ignition and the ground recommended that we yaw right 10 degrees.

SPEAKER   You had the rendezvous radar on?

ALDRIN   The rendezvous radar was on, not through the computer, but through its own AUTO track.

ARMSTRONG   We did not have the radar data feeding to the computer in the LGC position; but, apparently, if you have it in AUTO track, there s some requirement on the computer time. This is the way we ve been doing it in all simulations. It was agreed on. We were in SLEW. Prior to this time, we d been in AUTO track until such time as we started to lose lock in the pitchover. Then we went to SLEW, isn t that right?

ALDRIN   Are you talking about the program alarms during the descent? We ve passed the point of having the rendezvous radar in AUTO. We d switched it over to SLEW at that point.

ARMSTRONG   We were in SLEW with the circuit breakers in. Radar was turned on, but it was in SLEW. In the early phases of P64, I did find time to go out of AUTO control and check the manual control in both pitch and yaw and found its response to be satisfactory. I zeroed the error needles and went back into AUTO. I continued the descent in AUTO. At that point, we proceeded on the flashing 64 and obtained the LPD availability, but we did not use it because we really weren t looking outside the cockpit during this phase. As we approached the 1500-foot point, the program alarm seemed to be settling down and we committed ourselves to continue. We could see the landing area and the point at which the LPD was pointing, which was indicating we were landing just short of a large rocky crater surrounded with the large boulder field with very large rocks covering a high percentage of the surface. I initially felt that that might be a good landing area if we could stop short of that crater, because it would have more scientific value to be close to a large crater. Continuing to monitor LPD, it became obvious that I could not stop short enough to find a safe landing area.

9.2.25 Manual control/pitchover

ARMSTRONG   We then went into MANUAL and pitched the vehicle over to approximately zero pitch and continued. I was in the 20 to 30-ft/sec horizontal-velocity region when crossing the top of the crater and the boulder field. I then proceeded to look for a satisfactory landing area and the one chosen was a relatively smooth area between some sizeable craters and a ray-type boulder field. I first noticed that we were, in fact, disturbing the dust on the surface when we were at something less than 100 feet; we were beginning to get a transparent sheet of moving dust that obscured visibility a little bit. As we got lower, the visibility continued to decrease. I don t think that the altitude determination was severely hurt by this blowing dust, but the thing that was confusing to me was that it was hard to pick out what your lateral and down range velocities were, because you were seeing a lot of moving dust that you had to look through to pick up the stationary rocks and base your translational velocity decisions on that. I found that to be quite difficult. I spent more time trying to arrest translational velocities than I thought would be necessary. As we got below 30 feet or so, I had selected the final touchdown area. For some reason that I am not sure of, we started to pick up left translational velocity and a backward velocity. That s the thing that I certainly didn t want to do, because you don t like to be going backwards, unable to see where you re going. So I arrested the backward rate with some possibly spastic control motions, but I was unable to stop the left translational rate. As we approached the ground, I still had a left translational rate which made me reluctant to shut the engine off while I still had that rate. I was also reluctant to slow down my descent rate anymore than it was or stop because we were close to running out of fuel. We were hitting our abort limit.

9.2.28
Touchdown

ARMSTRONG   We continued to touchdown with a slight left translation. I couldn t precisely determine touchdown. Buzz called lunar contact, but I never saw the lunar contact lights.

ALDRIN   I called contact light.

ARMSTRONG   I m sure you did, but I didn t hear it, nor did I see it. I heard you say something about contact, and I was spring loaded to the stop engine position, but I really don t know whether we had actually touched prior to contact or whether the engine off signal was before contact. In any case, the engine shutdown was not very high above the surface. The touchdown itself was relatively smooth; there was no tendency toward tipping over that I could feel. It just settled down like a helicopter on the ground and landed.

ALDRIN   We had a little right drift, and then, I guess just before touchdown, we drifted left.

ARMSTRONG   I think I was probably over controlling a little bit in lateral. I was confused somewhat in that I couldn t really determine what my lateral velocities were due to the dust obscuration of the surface. I could see rocks and craters through this blowing dust. It was my intention to try and pick up a landing spot prior to the 100-foot mark and then pick out an area just beyond it such that I could keep my eyes on that all the way down through the descent and final touchdown. I wouldn t, in fact, be looking at the place I was going to land; I would be looking at a place just in front of it. That worked pretty well, but I was surprised that I had as much trouble as I did in determining translational velocities. I don t think I did a very good job of flying the vehicle smoothly in that time period. I felt that I was a little bit erratic.

ALDRIN   I was feeding data to him all the time. I don t know what he was doing with it, but that was raw computer data.

ARMSTRONG   The computer data seemed to be pretty good information, and I would say that my visual perception of both altitude and altitude rate was not as good as I thought it was going to be. In other words, I was a little more dependent on the information. I think I probably could have made a satisfactory determination of altitude and altitude rate by eye alone, but it wasn t as good as I thought it was going to be, and

I think that it s not nearly so good as it is here on Earth.

ALDRIN   I got the impression by just glimpsing out that we were at the altitude of seeing the shadow. Shortly after that, the horizon tended to be obscured by a tan haze. This may have been just an impression of looking down at a 45-degree angle. The depth of the material being kicked up seemed to be fairly shallow. In other words, it was scooting along the surface, but since particles were being picked up and moved along the surface, you could see little rocks or little protuberances coming through this, so you knew that it was solid there. It wasn t obscured to that point, but it did tend to mask out your ability to detect motion because there was so much motion of things moving out. There were these few little islands that were stationary. If you could sort that out and fix on those, then you could tend to get the impression of being stationary. But it was quite difficult to do.

ARMSTRONG   It was a little bit like landing an airplane when there s a real thin layer of ground fog, and you can see things through the fog. However, all this fog was moving at a great rate which was a little bit confusing.

ALDRIN   I would think that it would be natural looking out the left window and seeing this moving this way that you would get the impression of moving to the right, and you counteract by going to the left, which is how we touched down.

ARMSTRONG   Since we were moving left, we were yawed slightly to the left so I could get a good view of where we were going. I think we were yawed 13 degrees left; and, consequently, the shadow was not visible to me as it was behind the panel, but Buzz could see it. Then I saw it in the final phases of descent. I saw the shadow come into view, and it was a very good silhouette of the LM at the time I saw it. It was probably a couple of hundred feet out in front of the LM on the surface. This is clearly a useful tool, but I just didn t get to observe it very long.

ALDRIN   Here s a log entry: 46 seconds, 300 feet, 4 seconds after the next minute. Watch your shadow, and at 16 seconds, 220 feet. So I would estimate that I called out that shadow business at around 260 feet, and it was certainly large at that point. I would have said that at 260 feet the shadow would have been way the hell and gone out there, but it wasn t. It was a good-size vehicle. I could tell that we had our gear down and that we had an ascent and a descent stage. Had I looked out sooner, I m sure I could have seen something identified as a shadow at 400 feet; maybe higher, I don t know. But anyway, at this altitude, it was usable. Since the ground is moving away, it might be of some aid. But of course, you have to have it out your window.

9.2.23 LPD altitude

ARMSTRONG   The LPD was not used until we were below 2500 feet, and it was followed for some number of computation cycles. The landing point moved downrange with time as evidenced by successive LPD readings.

FCOD REP.   Do you recall when you proceeded?

ARMSTRONG   It was very shortly after we were going into P64.

ALDRIN   We got P64 at 41 minutes 35 seconds; then you went MANUAL, ATTITUDE CONTROL.

ARMSTRONG   I can t say whether that was before or after proceeding.

ALDRIN   It wasn t too long after that, 41:35-P64, 42:05-manual attitude control is good, 42:17-program alarm. What I m wondering is did the proceed have anything to do with maybe generating some more activity which would cause the program alarm? We weren t in 1668 at that point.

ARMSTRONG   I have no recollection of that area.

10.0 LUNAR SURFACE

ARMSTRONG   The post planning checklist went as planned, and venting was initiated in the OX tank almost immediately - even before the point on the timeline called for it. We ran the OX pressure down to the 40 to 50-psi range and opened the fuel vent and vented it down the same amount. We then closed it off after which time Houston became concerned with tank pressures and asked us to reopen it again, although at the time we were reading relatively low tank pressures. I think the OX built up slightly over 50 at the time, but they were apparently reading a higher value, and I assume perhaps that they had an error in their signal from that tank pressure. In any case, we opened both tanks after that and let them bleed down to about 15 psi, which was probably the stable condition of the vapor pressure in the tank at that point.

ALDRIN   After touchdown, we got a GO for T-1 and then we proceeded to enter P68 and recorded the latitude and longitude and altitude. We then proceeded out of that

and reset the stop button and entered P12 for T-2. At this point, I think that a little shuffling in the data cards might prevent someone from making the same error that I did in loading the tape for ascent. On the data card, we ve got the PDI pad, which is referred to somewhat during descent. It has PDI aborts on it with a NO PDI plus 12 abort on the right side. I think that the NO PDI plus 12 abort would be better placed on the back of this altitude card because, once you ignite, you re through with that NO PDI plus 12 abort and you ought to get it out of there. In its place, I think the T-2 abort pad should be on the data card because when I started to load P12 with NOUN 33 (the TIG for this T-2 abort, which is PDI plus 23), I  loaded the TIG for the NO PDI plus 12 abort, and the ground caught me on it and said, You loaded R-2 wrong. Instead of loading 10254 29, I loaded 10244 27. Now, the two are pretty close and they both say TIG NOUN 33. So, I think if we can get that one abort (NO PDI plus 12 abort) out of there and put the other one in its place, it ll save someone from coming up with the same sort of thing. We got remote control back to ATTITUDE HOLD and AGS, OFF and then cycled the Parker valves again. After having seen the erroneous talkback indications, I was expecting that when I put the feeds to close they would indicate open momentarily. The crossfeed was cycled again to close. I turned the camera off and proceeded with the switch configuration. Cycling the CWEA circuit breaker did, in fact, turn off the descent Reg warning lights. We read the sine and cosine out of AGS (and I m sure they copied that down) and went immediately into recording the AGS gyro coefficients. I m sure that these were the same numbers that we finished up with, but it might be a good time to check. Then we went to cabin on the regulators and took our helmets and gloves off. Then we started in with the initial gravity alignment. I don t have the first NOUN 04 that we got. I didn t record that one, but it was fairly large. I ve seen so many of them in simulation that I just can t recall what that number was.

ARMSTRONG   I am sure that it was recorded on the ground.

ALDRIN   After the recycle, it was 00001. We asked them about recycling and they said affirmative. Got a star-angle difference of 00015 and some torquing angles, which showed a fairly good change. I guess the pitch is the one you re concerned with. That s about the Y-axis; that was 0.1. I don t know whether the sine of that agrees at all with the approximation of 0.1 that we got in the Sun check. We didn t torque those angles. The ideal was to get a gravity direction and then to do a two-star alignment and look at the torquing angles after the two-star check which would then give an indication as to what the drift had been since the last alignment. The initial gravity alignment, combined with the two-star alignment, would produce a new location of the landing site. Had we landed straight ahead, my intent was to use Rigel in the left detent number 6 and Capella in the right detent. The 13-degree yaw moved Capella out of the right-rear detent, but Rigel was in good shape there. That s the one I used first. I then selected NAVI in number 4 detent, the right rear, and that wasn t particularly satisfactory. It was quite dim and it took a good bit longer than I had hoped to get the marks on that. I can t comment particularly on the star-angle difference other than it was a little disappointing in that it was 00009. Torquing angles we have recorded, and we did torque. The latitude and longitude - we d have to listen to the guidance people as to just what that did to our possible touchdown point. It seemed to me, when we finished that, we were just about on schedule, maybe a little bit behind so we proceeded into the option 3 which was gravity plus one star; because Rigel had been so good, I used that one again. The gravity alignment seemed to be quite consistent.

The first time we did the gravity alignment on option 3, it came up with 00000, our star-angle difference on the gravity plus one star, which indicates an error in that gravity measurement and star measurement was 00000. I know we had the torquing angle recorded on that also. The azimuth is very large - 0.2 degrees. We received a GO for T-3. In the vicinity of loading times for a T-2 abort, I noticed that the mission timer wasn t working. It was frozen; it just stopped.

ARMSTRONG   No, it didn t just stop.

ALDRIN   Yes, it had gone to 900 hours.

ARMSTRONG   900 and some hours. I couldn t correlate the minutes and the seconds with any particular previous event.

ALDRIN   Yes. 903:34:47. I don t know what time that relates to. Obviously, the 9 digit changed. It might have stopped, but it was static at that point. We ran through the circuit breakers but couldn t seem to get it moving again. The ground suggested that we turn it off which we did, but when we turned it back on we got all nines. You could change the last digit with the digit sequencer. We turned it off for a while and turned it back on again and it worked after that. We gave them an E memory dump; got a new ascent pad or the CSI pad, for T-3. We then proceeded on with the option 3 alignment. Continuing through the checklist, looking at switch settings, and circuit breaker cards, we found ourselves 10 minutes to go and essentially up on the checklist. At that point, we had to start pressurizing the APS if we were going to launch, so we read through the remainder of the simulated countdown and decided that there wasn t any point in sticking with that timeline any further. So we terminated the simulated countdown and went to the initial power-down sequence. We had discussed among ourselves the possibility of evaluating, during this first 2 hours, whether we wanted to go on with the rest period that was scheduled or to proceed with the EVA preparation. I think we had concluded before the end of the simulated power-down that we would like to go ahead with the EVA and it was sometime in here that Neil called to ground and let them know that.

ARMSTRONG   There were two factors that we thought might influence that decision. One was the spacecraft systems and any abnormalities that we might have that we d want to work on, and the second was our adaptation to 1/6g and whether we thought more time in 1/6g before starting the EVA would be advantageous or disadvantageous at that point. Basically, my personal feeling was that the adaptation to 1/6g was very rapid and was very pleasant, easy to work in, and I thought at the time that we were ready to go right ahead into the surface work and recommended that.

ALDRIN   Now, we estimated EVA at 8 o clock. I think that was a little optimistic. The ground recognized that, because they said, Do you mean beginning of PREP or beginning hatch opening? And all during this time, we could tell that Mike was kept busy each pass, doing P22 s trying - to find where we were.

10.8 HORIZON, SIGHTING, APPEARANCE

ARMSTRONG   The things that seem worthy of comment here are observations from the window prior to lunar surface work. We were in a relatively smooth area covered with craters varying from up to perhaps 100 feet in the near vicinity down to less than a foot, with density inversely proportional to the size of the crater. The smaller they were, the more there were of them. The ground mass was very fine silt, and there were a lot of rocks of all sizes, angularities, and types in the area. Our immediate area was relatively free of large rocks. Several hundred feet to our right there was a significant boulder field, an array of boulders, essentially, that had many boulders greater than 1 or 2 feet in size. We never were able to get into that area to look at those rocks in detail.

ALDRIN   Distances are deceiving. When we looked at this fairly large boulder field off to the right, it didn t look very far away at all before we went out. Of course, once we got out, we wandered as far as seemed appropriate. Of course, we never came close to this particular field. What really impressed me was the difference in distances. After we were back in again looking out at the flag, the television, and the experiments, they looked as though they were right outside the window. In fact, on the surface, we had moved them a reasonable distance away. So I think distance judgment is not too good on first setting down. The tendency is to think that things are a good bit closer than they actually are. This says they are probably a good bit larger than what we might have initially estimated.

10.10 COLORS AND SHADING OF LUNAR SURFACE FEATURES

ARMSTRONG   Probably the most surprising thing to me, even though I guess we suspected a certain amount of this, was the light and color observations of the surface. The down-Sun area was extremely bright. It appeared to be a light tan in color, and you could see into the washout region reasonably well. Detail was obscured somewhat by the washout, but not badly. As you proceeded back toward cross-Sun, brightness diminished, and the color started to fade, and it began to be more gray. As we looked back as far as we could from the LM windows, the color on the surface was actually a darker gray. I d say not completely without color, but most of the tan had disappeared as we got back into that area, and we were looking at relatively dark gray. In the shadow, it was very dark. We could see into the shadows, but it was difficult.

ALDRIN   We could see very small gradations in color that were the result of very small topographical changes.

ARMSTRONG   Of course, when we actually looked at the material, particularly the silt, up close it did, in fact, turn out to be sort of charcoal gray or the color of a graded lead pencil. When you re actually faced with trying to interpret this kind of color and that light reflectivity, it is amazing.

ALDRIN   When illuminated, it did have a gray appearance, very light gray.

ARMSTRONG   Wouldn t you say it is something like the color of that wall? It isn t very far away from what it looked like. Yet when you look at it close, it s a very peculiar phenomenon.

<u>10.15
PREPARATION
FOR EGRESS</u>

ARMSTRONG   Now, a preliminary comment has to do with the longer time that it took than during our simulations. It is attributable to the fact that when you do simulations of EVA PREP you have a clean cockpit and you have all the things that you re going to use there in the cockpit with you and nothing else. In reality, you have a lot of checklists, data, food packages, stowage places filled with odds and ends, binoculars, stop watches, and assorted things, each of which you feel obliged to evaluate as to whether its stowage position is satisfactory for EVA and whether you might want to change anything from the preflight plans. For example, our mission timer was out, and we decided we had better leave one wristwatch inside in case it got damaged. We would have at least one working watch to back up the mission timer or to use in place of the mission timer, in case we could not get it going again. All these items took a little bit of time, a little bit of discussion, which never showed up in any of our EVA PREP s on the ground, really accounted for the better part of an hour of additional time. Our view of EVA PREP was that we were not trying to meet a time schedule. We were just trying to do each item and do it right sequentially and not worry about the time. Well, the result was, a lot of additional time used there. I don t think that s wrong. I just think in future planning you are probably better off adding time for these kinds of things.

ALDRIN   No matter how many times you run through an EVA PREP, to the best of the instructor s ability to put things in a logical sequence, when you re faced with doing these things, there is a natural tendency to deviate somewhat from the printed sequence that you have. It s a rather complex operation. Nobody writes a checklist to tell you in the morning when you get up all the sequences you go through to put your clothes on, brush your teeth, shave, and all that. If you had one setting there, you wouldn t follow it the same every day. You would make small deviations just based upon what seems appropriate at that time. It is a very difficult thing to build a checklist for.

ARMSTRONG   We shouldn t imply that the EVA preparation checklist wasn t good and adequate. We did, in fact, follow it pretty much to the letter just the way we had done during training exercises. That is, the hook ups, and where we put the equipment, and the checks were done precisely as per our checklist. And it was very good. I don t have any complaints about that at all. It s these other little things that you don t think about and didn t consider that took more time than we thought. There was one control on the PLSS that surprised us. I don t know if it was different from the trainers or the flight PLSS s at the time we were looking at them or not, but there was a press-to-test knob of some sort that neither one of us could correctly identify as to function. At this time, we aren t really quite sure what it does.

ALDRIN   It was a thumb depress button that seemed to go in somewhere as if it was relieving some pressure from something. I can t remember ever having seen that before. It protruded out toward your back and looked as if it might come fairly close to riding on the back of the suit.

ARMSTRONG   We both thought we knew the EMU very well and knew every function and how it operated. But it turned out we were wrong. It was something that we hadn t learned there, and if it had been there before, somehow it escaped us. It took a little time to discuss that, and we proceeded.

ALDRIN   Mounting the 16-mm camera and the two universal brackets, one on the mirror mount and the other on the crash bar, went pretty much the way we had planned it to go. The two brackets with the enlarged knobs helped out tremendously in that I was able to tighten them down to a much greater degree than I had any of the training models. It gave me much greater confidence that the cameras would stay where I placed them and that there would be no problem with any camera banging into the window when we didn t want it to. The RCU camera brackets were difficult to tighten down. By tightening just as hard as we could, there was still a little bit of play in both of them. I think an improvement in that knob would be quite advantageous, so that it could be cinched down a little tighter. Perhaps the kind of knob that has edges that stick out so that you can get much higher torque on it would be a good thing to use.

ARMSTRONG   I think all the remainder of the EVA PREP went as per checklist.

ALDRIN   The heaters tested out. Both lights came on, pressures regulated at very close to 3.7. Then when it came time to unstow the hoses, the pressures had dropped down to just about zero.

ARMSTRONG   Yes. They were below 25.

ALDRIN   Overshoes went on quite easily. We put the antifog on as soon as we got the kit out instead of waiting until a little bit later. I think that maybe there were two things that brought that about. One was that we weren t really sure it was going to appear later in the checklist, and we wanted to make sure we had that. The other was, in training, we wanted to avoid as many activities as we could with the PLSS on our back because it was very uncomfortable doing any additional exercises in one g. We did find, however, that it was quite comfortable, even without the shoulder pads, to have the PLSS mounted on your back. The mass of it was not at all objectionable. It did require moving around methodically and very slowly to avoid banging into things - no getting around it. You just couldn t always tell what the back of the PLSS or the OPS might be in contact with at any particular time.

ARMSTRONG   As was reported, we broke one circuit breaker with the PLSS and we depressed two others, one on each side, sometime during the operation with the PLSS on the back. So that s an area that we still need to improve on to be able to have confidence that the integrity of the LM itself won t be jeopardized by the operation with the PLSS on the back.

ALDRIN   We had problems with this one particular electrical connector, the one that joins the RCU to the PLSS, ever since the first time we d ever seen it.

ARMSTRONG   It s about a 50-pin Bendix connector.

ALDRIN   It s just very difficult to get the thing positioned properly so that the three pins on the outside, the three little protuberances, will engage in the ramp so that, when you then twist, it ll cinch on in. That must have taken at least 10 minutes. The problem was not with mine, but in hooking up Neil s. I can t say that there was that much difference in the many times that I tried it unsuccessfully and the one time it did go in correctly. It appeared to be squared away each time.

ARMSTRONG   This is not because we didn t understand the problem. We had had trouble with that connector for 2 years or more. We d always complained about it. It had never been redesigned, and it was usually ascribed to the fact that all the training models were old and gouged, and so on. But when we looked at the flight units during CCFF on the EMU, it turned out that they were still difficult. We accepted the fact that by being very careful with that connector we could, in fact, connect and disconnect it satisfactorily. We did that in the lab at the Cape. We had a little bit of difficulty with it there. When we got on the lunar surface, it was the same problem. It took us at least 10 minutes each to mate those connectors. It s the big electrical cable from the RCU to the PLSS. It attaches at the PLSS end. It s our recommendation that it s a sufficiently serious problem that we can t afford to jeopardize the success of an EVA on that connector. And that s right now what we re betting. It began to look like we never would get those connectors made on the surface. We just have to improve that.

ALDRIN   Connecting up the straps went quite smoothly. The initial COMM check out on the audio panel and the various communications checks that we made in the FM mode all seemed to go quite well, until we started switching the PLSS modes. For a while, we ascribed some of the difficulty perhaps to the antenna being stowed. So we unstowed Neil s, and that didn t help immediately. A little later, it seemed to help out, but then we got back into about the same problem, so I stowed his antenna. There didn t seem to be any particular rhyme or reason to when we did appear to have good COMM and when we didn t.

ARMSTRONG   It suffices to say that we never did understand what was required to enable good COMM while we were inside the cockpit, relaying through the PLSS s. We had it part of the time, and we didn t part of the time. We tried a lot of various options, and they just weren t universally successful. But we were able to have adequate COMM to enable us to continue. I think, once outside, we really didn t have any appreciable COMM problems at all. It seemed to work well.

10.20
DEPRESSURIZATION

ARMSTRONG   This was one area of flight preparation that was never completely performed on the ground. In the chamber, the PLSS s were left on the engine cover and we never put them on our backs because of their weight, and the possibility of jeopardizing the integrity of the LM. So the COMM was operated, and the connections were made, but the depressurization sequence with the PLSS s on the backs was never completed. The times when we actually operated the PLSS was done always in the chamber and never done with the LM systems operable. So two things were new to us. One was that it took a very long time to depressurize the LM through the bacteria filter with the PLSS adding gases to the cockpit environment and the water boiler operation or something adding some cabin pressure. The second was that we weren t familiar with how long it would take to start a sublimator in this condition. It seemed to take a very long time to get through this sequence of getting the cabin pressure down to the point where we could open the hatch, getting the water turned on in the PLSS, getting the ice cake to form on the sublimator, and getting the water alarm flag to clear so that we could continue. It seemed like it took us about a half hour to get through this depressurization sequence. And it was one that we had never duplicated on the ground. Well, in retrospect, it all seemed to work okay, it was just that we weren t used to spending all that time standing around waiting.

10.21 OPENING
OF HATCH

ALDRIN   Well, there s a step verifying PGA pressure above 4.5, and decaying slowly. And it did that. It decayed slowly, and the cabin stayed at around I psi. We had to get that down before we could open the hatch, it appeared to me. We were just waiting there between those steps of PGA pressure and cabin pressure coming down, and opening the hatch. And we didn t really want to go and open the overhead hatch. We like to open only one of them, and leave the other one the way it s been. When the hatch was finally opened, it took an initial tug on it, and it appeared to bend. The whole hatch as it opened on the far side came toward me. As soon as it broke the seal, it appeared as though I could see some small particles rushing out. Then, of course, the hatch came open and gave us a more complete vacuum. Then we went to opening the water. It seems to me that, if there is that delay to get rid of the pressure, maybe one could go ahead and open up the water ahead of time before you actually get it down to the point where the hatch is open. Maybe that would compound the problem. Once the water window did clear, it seemed that the cooling was noticeable almost immediately.

10.22 FINAL SYSTEMS STATUS

ARMSTRONG   The final system status was without problems.

10.23 LM EGRESS

ARMSTRONG   I guess the most important thing here with respect to the egress through the hatch and the work on the ladder and the platform is that our simulation work in both the tank and in the airplane was a reasonably accurate simulation. They were adequate to learn to do the job and we didn t have any big surprises in that area. The things that we d learned about body positioning, arching the back, clearances required, and one person helping another and so on worked just like the real case. There weren t any difficulties in movement through the hatch or with stability on the porch. After getting onto the porch, I came back into the LM and went up around the Z-27 corner, made sure that was as expected and it was. I returned to the porch, got on the ladder, discarded our duffle bag with arm rest and OPS pallets, released the MESA without any difficulty, and descended the ladder just as expected. The first step was pretty high; 3 to 3 1/2 feet. So the initial test was to see if we would have any trouble getting back on the first step. There were no difficulties, so we proceeded with the planned activities. The work and effort required to go up and down the ladder and in through the hatch are not objectionable enough that they need be worried about. Going up the ladder and going through the hatch are not high-workload items. They are items that require some caution and practice. I had it a good bit easier than Buzz did because he had to go through the hatch and around the corner by himself.

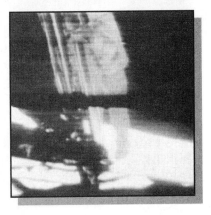

ALDRIN   Once I had my feet and posterior out the hatch, Neil was in good position, as good as I was to help me move out, by just observing the profile of the PLSS as it matched with the hatch opening.

ARMSTRONG   The two-man operation is good because all the help that each man can give the other one is money in the bank.

ALDRIN   I think the first man moving out has a little bit more difficulty because the second man has to be back behind the hatch and has to try to move it out of the way. So you have the tendency to be more over to your side away from the hatch and anything you are contacting was usually on your side, your edge of the lower part of the DSKY table.

ARMSTRONG   There weren t any temperature effects noted in the egress or ladder. Nothing felt hot or cold or had any temperature effects at all that I was aware of.

ALDRIN   The platform itself afforded a more-than-adequate position to transition from going out the hatch to getting on the ladder. The initial step is a little bit difficult to see. When I got to the first one, I was glad to have you tell me about where my feet were relative to that first step so that I didn t have to make a conscious effort to look around to the side or underneath. What I am getting at is that operations on the platform can be carried out without concern about losing your balance and falling off. There is plenty of area up there to stand on the step and do any manipulating that might be required. There are alternate ways of bringing things up, other than by the LEC. I think there is promise of being able to bring things up over the side; straight up, versus making use of the LEC. We didn t have the opportunity to exercise those.

In gravity fields, I would have come up with something closer to one-tenth, just by judging the difference in weight and feel of things in the way the masses behaved one to six. In the behavior of objects, it gives you the impression that there is a much greater difference. In my maneuvering, there didn t seem to be anything like a factor of 6 difference. It would appear as though the gravity difference was much less. What I m saying is that it seems the human can adapt himself to this quite easily. It also appears that objects can be handled easier in 1/6g than we had anticipated. In maneuvering the objects around, they do have a certain mass. When they get going in a direction, they will keep going that way. This was evidenced when the objects were coming in the hatch on the LEC; they were fairly easy to manage, but you had to take your time in handling them.

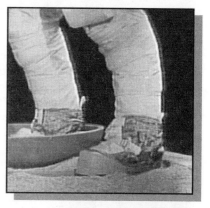

10.24 ADEQUACY OF HARDWARE AND PROCEDURES

ALDRIN   The initial LEC operation of lowering the camera seemed to work fairly well. It appeared as though you might have been pulling on the wrong strap at first; however, we rectified that without any particular trouble.

ARMSTRONG   Initially, I had a bit of difficulty. I was not trying to get the camera up or down at that point; I was trying to pull the slack out of the line and make both straps taut. For some reason or other, it was hung up, and I had some difficulty getting the slack out of the lines. Once having done that, it came down very nicely. Here we changed the flight plan somewhat and got the camera down before doing the contingency sample. I wanted to get that camera down and hooked up while I was over there in the shadow, because to do the contingency sample, I was going to have to stow the LEC and go over into the area out of the shadow. Since I wanted to do it on the right side where the camera was mounted, I was going to have to make a trip of about 10 or 15 feet before I started the contingency sample. That s the reason we changed the order. The operation of the suit, in general, was very pleasant. There was very little hindrance to mobility, with

the exception of going down to the surface to pick things up with your hands which was a very difficult thing to do. As far as walking around and getting from one place to another, the suit offered very little impediment to that kind of progress. It was, in general, a pleasant operation.

ARMSTRONG   Thermal loads in the suit were not bad at all; I ran on minimum flow almost the entire time. Buzz found a higher flow to be desirable. This was consistent with our individual preflight experience. I didn t notice any temperature thermal differences in and out of the shadow. There were significant light differences and visibility changes but no thermal differences. The only temperature problem I had (and Buzz didn t have this problem) was with the gloves. I did not wear inner gloves. I chose to go without the inner liners in the gloves, and my hands were a little warm and very wet all the time. They got very damp and clammy inside the gloves. I found that this problem degraded my ability to handle objects and to get firm grips on things.

ALDRIN   I had cooler levels set on the diverter valve, because it just seemed to be comfortably pleasant that way. In retrospect, it appears that this leads toward a higher consumption of water. I wasn t fully aware that when you are on higher flow, you are going to be pumping more water overboard. It was not clear to me preflight that it did have that effect on your water consumption. I certainly could have operated at lower levels much sooner without overheating. In confirmation of Neil s findings, I didn t experience any hot or even warm spots in the suit. I didn t wear any inner gloves, either, in my desire to get a better feel through the gloves. During the donning, I did not have the wristlets on. I thought that the LCG extending down far enough into the wrist would be adequate. If I had to repeat this effort, I would put the wristlets on,

because once I was in the gloves and I started moving them around, I did find that it was rubbing a small amount on the wrist. I thought that it might get to be more annoying than it actually turned out to be, but looking back, I would have preferred having those wristlets on.

ARMSTRONG   With respect to work on the surface, the 1/6 gravity was, in general, a pleasant environment in which to work, and the adaptation to movement was not difficult. I felt it was quite natural. Buzz had the opportunity to look at more detailed aspects of it, a good bit more than I did, but, in general, we can say it was not difficult to work and accomplish tasks. I think certain exposure to 1/6g in training is worthwhile, but I don t think it needs to be pursued exhaustively in light of the ease of adaptation.

ALDRIN   Moving around is very natural. Some attention must be paid to the mass that you have in the suit and also to the mass of the PLSS that is on your back. I think we anticipated this adequately, and the fact that we did have a sizable mass mounted to the rear was not detrimental to moving around.

ARMSTRONG   Buzz did more in that area than I did. I would say that balance was not difficult; however, I did some fairly high jumps and found that there was a tendency to tip over backward on a high jump. One time I came close to falling and decided that was enough of that.

ALDRIN   There is no doubt that it was much easier to reach that neutral point by just leaning back slightly than it was leaning forward. I think the happy medium was to lean forward more than we did. It was more comfortable for us to stand erect than to lean forward to be at that absolute neutral point. The POGO tends to give you the impression that most of your moving around will be the result of toe pressures - that you will rock up on your toes and tend to push off. I did not really find this to be the case as much as I had anticipated. The 1/6g airplane is a very poor simulation of the lunar surface. There is excellent traction in the airplane, so you can t relate too much as to how the foot departs or what sort of resistance you need when you put your foot back down again. I didn t find that there was much of a slipping tendency on the surface in trying to put in sideways motions or stopping motions. It was quite natural as you began to apply a force to make a change in your momentum. I think you were able to tell just how much you could put in before you would approach any instability case. In general, it would take a couple of steps to make a good sideways change in motion and it would take two or three steps to come comfortably to a stable - stationary position from a fairly rapid forward movement. To get a

sustained pace evaluation, I would have had to have gone a good bit farther than I did. Before the flight, I felt that you might be able to sustain a fairly rapid pace comfortably. My impression now is that this was a little tiring on the legs. There was a rubbing in the suit somewhere in the knee joints and you had to keep moving the knees, even though they are very mobile in the suit. I felt that, as easy as things looked, a 1-mile trek was not going to be an easy thing. Just by having to move your muscles and your body in the suit, you would end up getting tired on any prolonged trek. Because the terrain varies a good bit relative to your ability to move over it, you always have to be alert to what is coming up next. On earth you only worry about one or two steps ahead; on the moon, you have to keep a good eye out four or five steps ahead. I think the one foot in front of another is a much better mode of locomotion than the more stilted kangaroo hop. You can do it, but it doesn t seem to offer any particular advantage. When your feet are on the surface, you can do fairly vigorous sideways movements such as leaning and swinging

your arms without a tendency to bounce yourself up off the surface and lose your traction. This was one experiment that was suggested and I found that you do tend to remain well rooted on the surface where you are, despite motions that you may have. I guess the best thing in carrying this further is to answer the questions that people may have about certain specifics.

ARMSTRONG   I went the farthest. While Buzz was returning from the EASEP, I went back to a big crater behind us. It was a crater that I d estimate to be 70 or 80 feet in diameter and 15 or 20 feet deep. I went back to take some pictures of that; it was between 200 and 300 feet from the LM. I ran there and ran back because I didn t want to spend much time doing that, but it was no trouble to make that kind of a trek - a couple of hundred feet or so. It just took a few minutes to lope back there, take those pictures, and then come back.

ALDRIN   I don t think there is such a thing as running. It s a lope and it s very hard to just walk. You break into this lope very soon as you begin to speed up.

ARMSTRONG   I can best describe a lope as having both feet off the ground at the same time, as opposed to walking where you have one foot on the ground at all times. In loping, you leave the ground with both feet and come down with one foot in a normal running fashion. It s not like an earth run here, because you are taking advantage of the low gravity.

ALDRIN   The difference there is that in a run, you think in terms of moving your feet rapidly to move fast, and you can t move your feet any more rapidly than the next time you come in contact with the surface. In general, you have to wait for that to occur.

ARMSTRONG   And you are waiting to come down. So the foot motion is actually fairly slow, but both feet are off the ground simultaneously. You can cover ground pretty well that way. It was fairly comfortable, but at the end of this trip, going out there and back, I was already feeling like I wanted to stop and rest a little. After about 500 feet of this loping with a 1-minute stop out there in the middle to take pictures, I was ready to slow down and rest. There were a lot of interesting areas within 500 feet or so to go and look at if we had had the time. It would have been interesting to take that time and go out and inspect them closely and get some pictures, but that was a luxury we didn t have.

ALDRIN   There were so many of them, it is the sort of thing you just cannot anticipate before flight. You can plan to some degree when you are on the surface, but until you get out and look around, you can t make your final decision as to what you are really going to do. Inside, you are only looking at perhaps 60 percent of the available panorama.

ALDRIN   We were supposedly in a nondescript area, but there was far more to investigate than we could ever hope to cover. We didn t even scratch the surface.

ARMSTRONG   I ll be interested in getting the pictures back and looking at them. I think you ll find that even though it is not a terribly rough area - it is basically a smooth area - operating around in any type of a vehicle is going to take some planning. The Moon has fairly steep slopes, deep holes, ridges, et cetera. I am sure that we can devise things that will do that, but it isn t going to be just any vehicle that will cover that kind of ground.

ALDRIN   It will be interesting to see just how soon you depart from the walking-return concept. I don t think you can stretch that too far. I wouldn t guess as to what that distance is; you could give some reasonable distance you could return on foot, but it isn t miles. When you talk about miles, you are talking about being out of sight of the LM.

ARMSTRONG   Another area that is not listed here is the stereo camera. I would like to make a couple of comments about that. The stereo camera worked fine. We had no problems with it; however, it was hard to operate. I found that the angle that I had to put my hand on the handle to pull it and the force that it took was excessive.

ALDRIN   The squeezing of the trigger?

ARMSTRONG   Yes. I found my hand getting tired very soon while taking pictures with that camera. It was wearing out my grip.

ALDRIN   Would you say that the angle was too horizontal?

ARMSTRONG   Yes.

ALDRIN   You would like to have had it sloped down more towards you.

ARMSTRONG Yes. It was requiring the wrist to be cocked down.

ALDRIN The initial opening up or deploying of it went quite smoothly. The extension of the handle and the opening up of the case was quite well engineered. Separating the cover, taking it off, cutting the film, and removing the cassette also went quite smoothly. I think that the big area for reengineering might be just a change in the angle the handle comes out. We might have to add a hinge or something like that to it. What about the height of the handle? That would probably not be too bad.

ARMSTRONG I think that probably was reasonable. The other problem we had with the camera was that it was falling over all the time. I think this was the result of a little bit of difficulty in figuring out the local vertical.

ALDRIN Yes.

ARMSTRONG You d set it down and think it was level, but apparently it wasn t, because the next time you looked it would be laying over on its side. Or you would bump it inadvertently while you were looking somewhere else and knock it over. I picked it up three different times off the surface and it s a major effort to get down to the surface to pick the thing up.

ALDRIN How d you do that? By going down on the knee?

ARMSTRONG On one occasion I got it with the knee, one time I got it with the tongs, and the last time I had something else in my hand like a scoop or something that I could lean on and go down and get it. In general, there were a lot of times that I wanted to get down closer to the surface for one reason or another. I wanted to get my hand down to the surface to pick up something. This was one thing that restricted us more than we d like. We really didn t have complete clearance to go put our knees on the surface any time we wanted. We thought the suit was qualified to do that in an emergency, but it wasn t planned as a normal operation. We didn t let ourselves settle to our knees a lot of times to get our hand on the surface. Now I think that is one thing that should be done more on future flights. We should clear that suit so that you could go down to your knees, and we should work more on being able to do things on the surface with your hands. That will make our time a lot more productive, and we will be less concerned about little inadvertent things that happen.

ALDRIN Now we can say we have the confidence to know that we could get back up from the surface. You might have to put your hand down into all this. The thing that discouraged me was the powdery nature of the surface and the way that it adhered to everything. I didn t see any real need in getting down. I had no concern about doing it. But I agree. I think if we need something on the suit to qualify it to do this, then we ought to go ahead and do that. If it doesn t, if it just requires looking at the suits that we brought back and saying that they re qualified for kneeling, we ought to do that.

ARMSTRONG If you have a grip on something like a scoop, or a stick to hold on, then there s no problem at all in getting back up. You can go right down and just push on your hand and push yourself right back up. It was easy the time I did it with the scoop in my hand. That s one thing that we hadn t done a lot in our simulations, and it would be a help, I think. Let s go on with ingress.

<u>10.35</u>
<u>PHOTOGRAPHY</u>

ARMSTRONG Photography through the Hasselblads on the RCU mounts was satisfactory. I did have some trouble installing the camera on the RCU mount. The opening to the slot as you first put the tongue in the groove was binding a bit, and I always had difficulty getting it started. I d never observed that problem on the ground, and I can t account for it.

ALDRIN   I took the first panorama out in front without having the camera mounted on the RCU, and it did not appear to be unnatural to do so. It s much easier to operate with it mounted; however, I didn t find that the weight of the camera was as much a hindrance to operation as preflight simulations indicated it would be. There is no doubt that having the mount frees you to operate both hands on other tasks. The handle is adequate to perform the job of pointing the camera. I don t think we took as many inadvertent pictures as some preflight simulations would have indicated. It seems as though, in all the simulations where we picked up the camera, we always managed to take pictures. I don t think that was the case in this mission as much as we thought it was going to be. We ll know if a number of the pictures taken are pointed at odd angles.

10.36 SOLAR
WIND
COLLECTOR
DEPLOYMENT

ALDRIN   I found that the shaft extended and locked back into position very easily. It folded out, deployed, and unrolled. I was able to hook it in the bottom catch without any undue shifting around. In putting it in the ground, it went down about 4 or 5 inches. It wasn t quite as stable as I would have liked it to have been, but it was adequate to hold it in a vertical position. I could make the adjustments so that it was perpendicular to the Sun. The shadow that was cast by the solar wind aborted a good check in the fact that you did have it mounted perpendicular to the Sun. So I think we got a very high degree of cross-sectional coverage. When we get to surface penetrations, later, it s going to be quite evident that once you go past a depth of 4 or 5 inches, the ground gets quite hard. However, I didn t get much of a cue to this at this point while installing the solar wind experiment.

10.37
TELEVISION

ARMSTRONG   The TV was operated as planned with no particular difficulties. The one thing that gave us more trouble than we expected was the TV cable; I kept getting my feet tangled up in it. It s a white cable and was easily observable for a while, but it soon picked up this black dust which blended it in with the terrain, and it seemed that I was forever getting my foot caught in it. Fortunately, Buzz was usually able to notice this and keep me untangled. Here was good justification for the two men helping each other. There was no question about that either; he was able to tell me which way to move my foot to keep out of trouble. We knew this might be a problem from our simulations, but there just was no way that we could avoid crossing back and forth across that cable. There was no camera location that could prevent a certain amount of traverse of this kind.

ALDRIN   Neil initially pulled out about 20 feet of cable and then I pulled out the rest of it. It seemed to reach a stop; it seemed to have a certain amount of resistance, and I thought that was the end of the cable. However, when I pulled normal to the opening, I found that I could then extract the cable to the point where I saw the black and white marks on it. The cable, being wound around the mounting inside the MESA, developed a set in it so that when it was lying on the surface in 1/6g, it continued to have a spiral set to it which would leave it sticking up from the surface 3 or 4 inches. It would be advantageous if we could get rid of that some way.

ARMSTRONG   Your foot is continually going underneath it as you walk, rather than over the top of it.

ALDRIN   One time when Neil did get the cable wrapped around his foot, the cable very neatly wrapped itself over the top of the tab on the back of the boot. That created a problem in disentanglement. I don t know whether it s worth moving that tab or not.

10.38 BULK SAMPLE OPERATIONS

ARMSTRONG   The bulk sample took longer than in the simulations because the area where the bulk sample was collected was significantly farther from the MESA table than the way we had done it in training. The MESA table was in deep shadow and collecting samples in that area was far less desirable than collecting them out there in the sunlight where we could see what we were doing. In addition, we were farther from the

exhaust plume and the contamination of the propellants. So I made a number of trips back and forth out in the sunlight and then carried the samples back over to the scale where the sample bag was mounted. I probably made 20 trips back and forth from sunlight to shade. It took a lot longer, but by doing it that way, I was able to pick up both a hard rock and ground mass in almost every scoopful. I tried to choose various types of hard rocks out there so that, if we never got to the documented sample, at least we would have a variety of types of hard rock in the bulk sample. This was at the cost of probably double the amount of time that we normally would take for the bulk sample.

ALDRIN   I want to inject a thought about spacecraft location in respect to lunar surface working location. Putting the area of the MESA in the shadow also put the cable in the shadow. The white cable, being covered with a little bit of this powdery stuff and being in the shadow, was very difficult to observe. Consideration should be given to keeping any cable or small object out in sunlight whenever possible. It leads one to think that if you re going to yaw one way or the other, it s preferable to put your working areas out into the sunlight.

ARMSTRONG   We ve discussed free-launch on a number of occasions and whether we wanted to yaw specifically for lighting at touchdown. There are obviously a lot of advantages, but I was very reluctant to do any fancy maneuvering on the first lunar touchdown for selected yaw for lighting considerations. I figured we d just take what we got and we paid for that later, because we had a lot of operations in the shadow during EVA that would have been easier had we had better lighting.

It s very easy to see in the shadows after you adapt for a little while. When you first come down the ladder, you re in the shadow. You can see everything perfectly; the LM, things on the ground. When you walk out into the sunlight and then back into the shadow, it takes a while to adapt.

ALDRIN   In the first part of the shadow, when you first move from the sunlight into the shadow, when the Sun is still shining on the helmet as you traverse cross-Sun, you ve got this reflection on your face. At this point, it s just about impossible to see anything in the shadow. As soon as you get your helmet into the shadow, you can begin to perceive things and to go through a dark adaptation process. Continually moving back and forth from sunlight into shadow should be avoided because it s going to cost you some time in perception ability.

ARMSTRONG   We ll start here with the flag installation. It went as planned except that the telescoping top rod could not be extended. Both Buzz and I operating together were unable to put enough force into extending the rod. It appeared to just be stuck and we gave up trying. So the flag was partially folded when we installed it on the flagstaff. I suspect that didn t show very much on television but our still photographs should show the result of that.

ALDRIN   Neither of us individually could extend it. We thought maybe we could extend the rod by both pulling, but then we didn t want to exert too much force because if it ever gave way, we d probably find ourselves off balance. I don t know how we ll ever find out what happened. I suspect this is just something that may in some way be due to thermal conditions or vacuum welding or something like that. It came out of its mount fairly easily. I thought we had a little bit of trouble with one of the pip pins there for a while. Generally, it was a straight-forward job to dismantle it.

ARMSTRONG   The flagstaff was pushed into the ground at a slight angle such that the c.g. of the overall unit would tend to be somewhat above the point at which the flagstaff was inserted in the lunar surface. That seemed to hold alright, but I noted later after getting back into the LM that the weight of the flag had rotated the entire unit about the flagpole axis such that the flag was no longer pointed in the same direction as it was originally. I suspect that the weight of the flagpole probably had shifted its position in the sand a little bit from the position where it had originally been installed.

ALDRIN   How far would you estimate you got it into the ground?

ARMSTRONG   Six to 8 inches was about as far as I could get it in.

ALDRIN   It was fairly easy to get it down the first 4 or 5 inches.

ARMSTRONG   It gets hard quickly.

**10.39 LM
INSPECTION**

ALDRIN   I don t think we noticed a thing that was abnormal. I guess the only thing that I made note of was the jet plume deflectors. The one on the right side as I was looking at the LM (which would make it the quad 1) appeared to be a bit more wrinkled than the one on quad 4. Of course, there s nothing to compare it with, because I d never seen them before. As a matter of fact, the first time we really saw them was when we looked out of the command module and got a pretty good idea of their structure.

ARMSTRONG   The only abnormality I noticed was (and it wasn t an abnormality) that the insulation had been thermally damaged and broken on the secondary struts of the forward leg.

ALDRIN   This is true in the rear, also.

ARMSTRONG   We didn t carefully check every secondary strut, but the primary struts didn t seem to be damaged.

ALDRIN   Yes, in the foot passage, it didn t appear to have suffered hardly at all. There was a sooting or darkening or carboning; I don t know what you call it. At least, I feel it was a deposit rather than just a baking or singeing of the material.

ARMSTRONG   We have some pictures of the struts.

ALDRIN   The part that had been melted, separated, and rolled back or peeled back on the secondary strut appeared as though it was a much more flimsy design then any other thermal covering on there. I don t think there is anything significant in the fact that part of the thermal coating that was higher up had separated, whereas the material lower down had not. I didn t notice anything peculiar about the vents. There didn t seem to be anything at all deposited on the surface from any of the vents underneath or from the oxidizer fuel vent up above.

ARMSTRONG   The most pronounced insulation damage was on the front plus Z strut. Its being in deep shadow obviated the possibility of getting a good close-up picture in that dark environment.

ALDRIN   I think the best pictures we got were of the minus Z strut.

ARMSTRONG   There was less damage than on the examples we looked at preflight. Just the very outer layers were penetrated.

ALDRIN   From what I could see of the probes, they had just bent or broken at the upper attach point. I didn t observe that they had any other fractures in them. One of them on the minus Y strut was sticking almost straight up.

ALDRIN   It was pretty substantially the metal case on the outside of it, and there weren t any thermal effects noted on it at all. The inner thermal coating was trying to protect something that was relatively fragile, the flag itself, however, there was no sign of degradation on the flag. I don t remember seeing the minus Z probe. I don t know; maybe it was there.

ARMSTRONG   I thought I remembered seeing all three probes. I think one was straight up and one had a V shape.

10.40 EASEP DEPLOYMENT

ALDRIN   Taking the cover off the lanyard was very easy. It pulled away and didn t seem to have any thermal or blast effects on it.

ALDRIN   Underneath the EASEP, the radar looked like it came through without any heat damage that I could tell. The lanyard underneath the thermal cover was in great shape. I didn t see any evidence of thermal effects. When it folded out, the doors went up even easier than the trainer. As the top door folded back, it didn t seem to fall into a detent and I tugged on it a couple of times. It looked like it was going to stay up there without any tendency to come back down again. In an effort to save some time, I elected to deploy both packages manually; I pulled out the seismometer a few inches, disengaged the hook, disconnected it from the top, and slid it out. I was unable to toss the lanyard over the side door to keep it out of the way, so it did come down from the boom and had a tendency to get in the way. The package itself was quite easy to manage. I had my left hand on the handle and moved the right hand around to support the weight as it slid off the rails. It was disengaged quite easily from the boom at the pip pin. I had it down on the surface, and then to get ample maneuvering room to get the retro-reflector down, I decided that I wanted to move the seismometer away. However, there happened to be a small crater right there, so I had to move it maybe 10 feet away and come back. Remember, it didn t seem to be a good place to set that seismometer down, other than right in front. It appeared to be in my way a little bit. In pulling out the laser package, I used the same technique, pulling out a few inches, then disconnecting the lanyard from the package itself, then pulling the string that was attached to the pip pin. In training sessions, I had pulled this one

rather slowly and firmly and had a few problems with the pip pin binding. The recommendation was to give it a fairly good jerk. When I did this, the wire ring that attached the cord itself to the pip pin sprung open. Either it was a welded joint that separated or thermal effects somehow weakened it; but it opened up and came loose from the pin. I was able to get the pin out by depressing the one side. Then by pushing it with my right hand and pushing it through, it came loose. Then I lowered it down to the surface and again it was quite easy to handle. The boom slid back in with no problem. I left the lanyards dangling out the bottom, pulled the retract lanyards, and the doors came back down and fitted together very nicely. The whole operation was quite smooth and I thought we got a little bit ahead in time in the deployment of these things. I picked up the two packages and we headed out to the minus Y strut looking for a relatively level area. Looking for level areas, I found it difficult in looking down at the surface and saying exactly what was level. I don t know what to attribute this to particularly. You don t have as good

a horizon definition as on the earth. When you look out to the side, you ve got a very flat area on the Moon. When you look out to the edges, you ve got varying slopes. I think it s further compounded by the fact that with 1/6g, and a center of mass displaced considerably aft and up from where it normally is, your physical cues of supporting your weight are different. The result was that it was just a little bit difficult to tell what was level and what was sloping, either to one side or up or down.

ARMSTRONG   You don t have as strong a gravity indication either, I don t think.

ALDRIN   Yes. It doesn t have as firm an orientation. That pretty well covers the deployment out to the site. In going through the numbers of pulling the little lanyards, every thing progressed as neat as can be. The handle deployed upward and rotated around, even though I wasn t able to see it fit into its slot. This is the maneuvering handle on the PSE. I might point out that the flight article was different in configuration than the training package, the difference being that you couldn t see when the handle was out and locked in its detent as well on the flight package as you could on the training package. Anyway, this worked out quite well. Orienting the package in azimuth was quite easy. The shadow of the gnomon stood out quite well in our session in the lab with the flight packages. We had had some concern as to just how well this shadow was going to stand out against this silver surface. However, all three of the pins in the gnomon were quite clear. I won t say they were a very crisp shadow, as there was a little bit of fuzziness to them, but it was quite easy to determine where the center of it was and get it orientated at the 45-degree mark. The big problem arose in trying to get the BB to settle down into the center of its little cup.

ALDRIN   It seemed to want to find a home away from me at about 11 o clock as I faced the package. I would try to push it down to get it to rotate around and it would move away from this position and start spinning around the outside. Try as I would to move it gradually away or push down on the package (away from where the bubble was) to get it to drift across, I was completely unsuccessful in getting the BB to find a home anywhere but along the perimeter. As I would bend down and look at this thing, it just appeared that this cup, instead of being concave, had somehow changed its shape and was convex. It didn t appear that there was any hope of the BB ever being anywhere but along the edge, so I visually tried to level it as best I could. As I indicated before, that wasn t too easy to do with any degree of confidence. Then I went to deploy the panels. One of the two retaining structures that should have fallen away when you right the package (both should fall down exposing the panels) failed. So I walked around the package and easily reached down with my finger and flicked it loose. It didn t require much force at all. When I deployed the panels, the left one came

out and deployed completely; then following another pull on the lanyard, the right one deployed. There was a certain amount of rocking motion and dancing around on the surface as the two deployed panels flung themselves around before finally settling down. During the process of doing this, I believe two of the four corners came in contact with the surface and picked up a light coating of surface material. I d say the triangle that was coated might have been 2 inches on one side and maybe I inch on the other - a very small triangle. So I don t think there was much degradation at all on the surfaces with that particular coating. I made one final inspection and, when I left it, the BB was still sitting on the edge. Neil came by with the camera to photograph it and he looked at it and found the BB was sitting right in the center of it. I have no explanation for that at all.

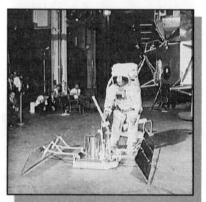

ARMSTRONG   It would have been nice to have a big rock table to set those packages on, but there wasn t any. The area where they were placed was a ridge between some shallow craters. I think we have reasonably good pictures of those ridges. They have this same kind of soil consistency as the surrounding area. The packages were in essentially soft material which allowed us to jiggle them down and get them reasonably well set into the sand, but there is no knowing whether they will stay there for a long period of time or might slowly settle.

ALDRIN   I think that they retained their present position pretty well. When I decided that I wanted to change the slope of the package one way or another, I found that I had difficulty in getting it to sink down a little more on one side. Even by scraping it back and forth, I couldn t seem to lower one edge as much as I would have liked to have.

ARMSTRONG   There was no difficulty in the laser reflector installation. It worked as we expected.

ALDRIN   Let s discuss the documented sample. We were obviously running out of time at the end of the EASEP deployment. We had limited time to conduct the documented sample. A figure of 10 minutes was used. I thought we might actually progress in a formal excursion and get something started anyway. As the box was opened, we got the report that they wanted two core tubes and it looked like that was probably going to take most of the time. While I proceeded to that - because that s essentially a one-man operation - Neil went around the backside of the LM and picked up what rocks he could identify, getting as wide a variety as possible. In unpacking the box with the core tubes, I was quite careful to try to identify where the caps were. In some simulations, we had misplaced them, they had dropped to the surface. I do think we need a better way of identifying the various packages that have this packing material wrapped around them, so that at a glance you d know what is inside a certain roll. In many cases, there is nothing in

it. In other cases, its got an environmental container in it, or its got the caps to the core tubes. In putting the extension handle on the core tube, the first one went on fairly cleanly and locked into position with a fairly high degree of confidence that it was not going to come out. I won t say that there was complete certainty that they were not going to come apart. I then picked up the hammer, went out into the vicinity of where the solar wind experiment was, and drove the first core tube into the ground. I pushed it in about 3 or 4 inches and then started tapping it with the hammer. I found that wasn t doing much at all in the way of making it penetrate further. I started beating on it harder and harder and I managed to get it into the ground maybe 2 inches more. I found that when I would hit it as hard as I could and let my hand that was steadying the tube release it, the tube appeared as though it were going to fall over. It didn t stay where it had been pounded in. This made it even harder because you couldn t back off and really let it have it. I don t know if we have any way of measuring the exact force or impact that was applied other than subjective. Maybe watching television would be some help. I was hammering it in about as hard as I felt I could safely do it. Unfortunately, we don t have any of the surfaces on the extension back to look at the impact. I was hitting it with the hammer to the point that I was putting significant dents in the top of it. I didn t find any resistance at all in retracting the core tube. It came up quite easily. On rotating it up to the inverted position to keep anything from coming out, I didn t find any tendency at all for the material to come out of the core tube. When I unscrewed the cutter, the surface seemed to separate again without any tendency for the material to flow or move. This meant that the consistency of this material, even though it looked to be about the same was a good bit different. If I had some very close surface material and shifted it a little, it would tend to move from one side to the other. At the

bottom of the core tube, I had the distinct impression, and it s just a descriptive phrase, that this was moist material. It was adhering or had the cohesive property that wet sand would have. Once it was separated from the cutter, there was no tendency at all for it to flake or to flow. I put the cap on, put it away, and then went to another area I would judge 10, maybe 15 feet away. I encountered about the same difficulty in driving the tube in. I imagine it went in about the same depth. It struck me that when I was removing this core tube from the extension handle, it was coming off. I had less confidence in initially putting the two together that they were going to stay together properly. When I was removing it, it appeared as though the end of the core tube that attaches to the extension handle had a tendency to come off. I had noted this earlier in some of the bench checks. When you screw the core tube in, if you aren t careful when you disengage it you re liable to disengage the cap on the other end. And the reason I m belaboring this particular point is because I understand that one of the ends did come off. I guess I can t be sure that it did not come off at the time of disengaging. Perhaps it could have come off in the box, but I don t believe they found the other end. So the assumption is that when it was taken off the extension handle, the other end came off with it. It doesn t appear as though the material spread around inside the box because none could be found, so it must have adhered pretty well. Did we get photos of both those areas?

ARMSTRONG   I did not get stereo pairs. I got one photograph of the second one. Well the first one to a high degree of confidence was right in the area of the SRC. We can identify its location pretty well by the photograph. The solar wind disengaged from its staff quite easily. When it rolled up, it had a tendency to sneak off to the side and crinkle on the edges. I spent some 20 to 30 seconds unrolling it and trying to get it to go up a little smoother. I then remembered that they really didn t care about exact neatness. All they wanted was the material back because they were going to cut it up in many pieces anyway.

So I bunched it together and it slid into its container fairly easily. In regard to the SRC height, we couldn t tell, due to the insulation, just what it was; but we gave the height of our ladder above the ground. The photographs would fill in the story there.

ALDRIN   It might be advisable to have some simple measuring device. It wouldn t take very much. Perhaps by the use of some marks you just make a judgment whether the distance between the 3 and the 4 is the same as between the 4 and the 5 or whatever the sequence might be. The SRC s worked as planned. The only difficulty that I encountered was in closing the boxes. Opening the second one, I felt, required a little more force than I had anticipated in just lifting up the lever lock.

ARMSTRONG   Closing the bulk sample box took a lot more strength than I had expected. It took just about everything I could do to close the document sample box. I was afraid I might have left the seal in the box. I don t think I did because, at the time, I thought I remembered clearly taking the seal off and throwing it away; but that s what it felt like. I inadvertently tried to close one with the seal in place at one time during training, and this was very much the same kind of situation. It took an inordinate amount of force. There s another difficulty in the fact that the gravity is so low that the box tends to slip around very easily. It feels very light; skids away from you. So, in addition to closing it, you have to hold it firmly down on the table. The table s not very rigid. It s quite flexible. So just holding the box securely enough in position to apply the high force on the sealing handles was some trouble.

QUESTION   Compare lunar versus Earth gravity.

ALDRIN   Subjectively comparing the weight of the boxes (following removing them from the spacecraft on the carrier), I would say closer to one to ten - just judging the differences in weight and feel of things and the way the masses behaved. One to six gives you the impression there is a much greater difference than that. Now in your own maneuvering around, it doesn t seem to be anything like a factor of six in the ease in being able to do things. It would appear as though the gravity difference was much less. What I m saying is that it looks like the human can adapt himself to this quite easily. It also appears as though the handling of objects is considerably easier in 1/6g, as we had anticipated. In maneuvering objects around, they do have a certain mass. When they get going in a direction, they will keep going that way as was evidenced when they were coming in the hatch on the LEC. They are fairly easy to manage, but you have to take your time in handling them.

10.44 LM INGRESS

ALDRIN   Stability and balance. Well, the first step up to the bottom rung no doubt is a pretty good step, though Neil tells me he got up to the third one.

ARMSTRONG   The third step.

ALDRIN   The capability exists to do a good bit more in terms of a vertical jump than certainly the POGO leads you to believe. There s no way to evaluate that in the airplane. The big problem in the POGO was that it just didn t seem to be able to bring you down with enough to bear so that your inertia would carry you as far as it s able to with good leg extension.

ARMSTRONG   The technique I used was one in which I did a deep knee bend with both legs and got my torso down absolutely as close to the foot pad as I could. I then sprang vertically up and guided myself with my hands by use of the handrails. That s how I got to the third step which I guess was easily 5 to 6 feet above the ground.

ALDRIN   The rungs of the ladder were not in any way dangerously slippery. Material on the bottom of your boots tended to cause them to slide back and forth.

ARMSTRONG   They were a little slippery.

ALDRIN   I think we have already mentioned the adequacy of the platform for other operations, that is, alternate ways of bringing things up. The hatch moved inward very easily. As I faced the hatch, I moved the camera from its position on the right side of the floor, up onto the Z-27 bulkhead. I had very little difficulty, again using the same technique that Neil used. About half way in, make concerted effort to arch your back to keep the PLSS down by keeping your belly down against the floor. This affords you the least profile going in. There didn t seem to be any exertion at all associated with raising yourself up and transitioning to a point where you can bring your knees on inside the cockpit, and then moving from a kneeling to an upright position. It all seemed to work quite smoothly. When there is a large bulk attached to you, you have to be careful. Once you get inside, before you start to turn around, you must make adequate allowance for all this material behind you.

ARMSTRONG   That was an interface problem. As a matter of caution, each person should be helping the other as much as possible. The first man in has the biggest problem, at least when he gets inside the cockpit. He has nobody to help him with clearance and I m sure he must use a good bit of caution.

ALDRIN   The LEC didn t seem to get in the way at all while I was getting in. We had the mirror available, but I don t think either of us found any particular use for it.

10.45 EQUIPMENT JETTISON

ARMSTRONG   I don t recall any difficulties with that.

ALDRIN   Well, this period was prolonged a bit to try to make as much use of the film remaining. I think we probably took more pictures than we should have in an effort to make sure that we covered each particular window as thoroughly as possible and with as wide a range of settings as we could before we proceeded to jettison the camera.

ARMSTRONG   I think the equipment jettison went well and as planned.

ALDRIN   We made an LiOH change at this point.

ARMSTRONG   We included the canister as a separate jettisonable item at this point, which we had planned to do before the EVA.

ALDRIN   We elected to leave the helmets on because at this point there was so much stuff rattling around inside the cabin that they would have added just one more bulky item. The primary canister change proceeded quite well to the point of inserting the new canister. I ran into a minor problem in getting it to rotate fully so that I could get the cover on. When it finally did seat itself in properly, I can t for sure identify what I did differently from the times when it didn t seem to rotate. That seemed to be what was stopping the cover from going on completely, the fact that when the canister was inserted I couldn t seem to rotate it as much as I thought it should have been rotated.

However, the canister container behind the ascent engine removed very easily, and we were able to jettison it without any problems. We didn t have any problem; I didn t notice you had any difficulty giving the packages the heave-ho. I think each PLSS bounced once on the porch before it went on down.

ARMSTRONG   Only one thing stayed on the porch. That was a small part of the left-hand-side storage container that did not make it off the porch onto the surface. That was the last item jettisoned. Concerning the LEC, I had neglected to lock one of the LEC hooks which normally wouldn t have caused any trouble. You would expect to proceed normally whether that was locked or not. However, for an unknown reason when I got the SRC about half way up, the Hasselblad pack just fell off. I can t account for that. I just took the pack on up and attached it, and ensured that it was locked when I put it on the SRC the second time. When it fell onto the surface, it was covered with surface material.

ALDRIN   I m sure there is a lot of inertia with any package like that and, with that low gravity, it tends to swing back and forth; and if there is some tendency to reach an unlocked position, it will.

ARMSTRONG   There was no problem because the ladder was right there, so I just leaned over and down to the ground and picked it up. I had the ladder to hold on to and then I could push myself right back up to a standing position.

ALDRIN   Did the film magazine hit the pad or drop right to the surface?

ARMSTRONG   I think it hit the surface clear of the pad, on the right side, which would be the spacecraft s left. I wasn t worried about the contingency sample because that was inside a bag. If anything was going to catch fire, it was going to be my whole suit because it was just covered with that stuff.

ARMSTRONG   The post-EVA checklist went very well. It was well planned and we went precisely by the preplanned route with possibly a few exceptions. They went very well and probably took about the same or a little more time than we expected. Of course, the time period that we took while we were waiting for the canister before starting the repressurization was comparably long. We had to put an eat period in there as I remember and took a lot of pictures.

ALDRIN   Well, there s no getting around it, it s another EVA PREP exercise. It s easier, but you still have to go through the same exercises such as pressure-integrity check, reading the cabin down, and configuring the ECS. I guess if you have two EVA s, it probably would be nicer to jettison your equipment at the beginning of the second one, rather than having to add another DEPRESS. I m not sure how they re planning to do this.

ARMSTRONG   There still was a full truckload of equipment inside that cockpit at the end of EVA. It s just a bunch of stuff, and I was glad that we were able to get rid of a lot of it and finish the jettison before we started our sleep period. With all that stuff in the cockpit, there s really no place left for people to relax.

ALDRIN   On P57 before lift-off, the Sun moved up in the field of view, as did all the rest of the stars. The Earth stayed the same. The Earth obscured the forward detent and the right detent. The Sun was now in the rear detent, and for some reason, it also obscured the left-rear detent, which was the one I was counting on using with Rigel. This was the one we had used before. I was quite surprised to discover this. The Sun was not within more than 15 degrees of the total field of view. It completely obscured the left-rear detent. It effectively left us two out of the six detents to pick stars from. Looking at those two detents, there weren t any stars near the center. The closer to the center of the detent you get the greater the accuracy is. The day before we had used Navi and it wasn t particularly bright. So I went back and now could use Capella, but it was fairly close to the edge of the field of view. So we did a gravity/one-star alignment and that first gravity alignment came up with 00010. VERB 32 gave us 00001. We used a sequence of marking that involved an onboard averaging of five successive cursive readings, followed by depressing the MARK button, and then five successive spiral readings that Neil would log down as I would read them off. Then he would average these up and we would put them in.

We d use either spiral or cursive first, whichever appeared to be convenient. I think this averaging technique worked out better than letting the computer do it, because it would have amounted to a considerable rotating of the spiral and cursive reticle field back and forth to make one spiral, then a cursive, then a spiral, and do a recycle. There is the option, however, to do one or the other. This was a REFSMMAT alignment. The torquing angles were fairly large, the star-angle difference was 00007 which preflight was the expected value of a two-star alignment. Torquing angles were very close to 0.7 in all three axes, which indicated that the platform did drift a fair amount during that time period. We then did the P22. I had hoped at that point to use the AGS to tell me where the command module was, but unfortunately we didn t update the AGS with the latest PGNS state vector so it wasn t giving us good range and range rate. I would recommend doing that, if anyone does a P22 in the future, because you can t use the PGNS to tell you what the range and range rate are. And you can t use the radar because it s not going to lock on until it gets to 400 miles. But the AGS gives you very good indications as you are approaching that range.

So we were a little misled and I thought we were still well out of range when we finally got the lock-on. You call up the program before the command module gets to 400 miles. It sits and waits; and, when it gets less than 400, it locks on automatically, and you see the signal strength grow and it starts to track. But it s in mode 2 so you don t see the needles doing anything; the cross pointers move, indicating it s got rate drive going as it s trying to keep up with it. Because we didn t want to run the tape meter into the stops, we left it in ALTITUDE/ ALTITUDE RATE. We really didn t have much of an indication that any good information was coming in, other than signal strength. I guess the ground got the data on the downlink. When it broke lock, I thought the command module was overhead and it had broken lock because of a maximum rate drive. The radar representative from RCA had indicated that the SPEC said it might break lock, but he didn t think it would

as it went over the zenith. But, because of the AGS indications, I thought that was what had caused the break-lock. Evidently, it had gone out the front field of view. It broke lock just a short time after the time given us by the ground for the zenith passage. So I was fully expecting it to acquire again. I don t think we had our AGS configured and the ground was not as helpful as they might have been had we run this sort of thing previously in simulations and had a bit more training on it. We started to do the P57 and realized that this would be too soon before lift-off. It seems to me we had a time period in which we were essentially standing by. We did an abbreviated RCS check. Because one of them was a cold fire check, we got all the red flags coming on. We did an AGS calibration, got the ascent pad; then, at about 45 or 50 minutes before lift-off, we called up the P57 again. We did a landing site option at the TIG of lift-off. The torquing angles between this alignment and the previous one were on the order of 0.09 degree maximum. The gravity alignment had an initial error of 0.00001 and on recycle had the same thing. I don t have logged down what the star-angle difference was, but it was probably on the order of 7 to 9, somewhere in there. It wasn t anything that made me jump up and down. But again it was measuring the difference between gravity and a star and, of course, doesn t really indicate how well you know the star position or how well you measured that, because it s relative to how well the gravity was measured. We had an update concerning the position to leave the radar for ascent. We were instructed by the ground not to turn the radar on during ascent and to leave it in SLEW. I think we left the circuit breakers out. This was to keep from overloading the computer, in a similar way to what had happened during the previous day during descent. I think that s unfortunate that we do have to deprive ourselves of one additional check for insertion confirmation. There was one more venting of the

descent tanks at insert - lift-off minus 30 minutes. I had the radar in SLEW and the circuit breakers off.

ARMSTRONG   I m quite sure they were off.

ALDRIN   Well, I didn t want us to use the tape meter in PGNS. Now that would have given us altitude and altitude rate out of the PGNS, right? So they didn t want to burden down the PGNS with doing that. Here I have on the circuit breaker card, leaving both radar circuit breakers open. We got the batteries on the line a little sooner because I think the ground thought that they might have cooled down a little bit more than their preflight information might have indicated. So we brought those on before TIG minus 30. Another change - we lifted off with the update link in VOICE BACKUP, brought the VHF ranging on at TIG minus 15, and pressurized the APS tanks. I guess it slipped my mind, perhaps Neil s too, that the Apollo 10 crew had noted that they saw very little decrease in the helium pressure. At first, it looked like we had about a 100-psi decrease in Tank 1 and zero decrease in Tank 2. That was probably the worst thing we could have seen because we figured that just one tank had pressurized. The ground was a little concerned about that. If they were not concerned, I wish that they had given us just a little bit more comforting thoughts at that particular time, because we hesitated at that point, at least I did, in doing some more of the switch configuration, waiting for a confirmation from them.

10.51 GENERAL LUNAR SURFACE FATIGUE

ARMSTRONG   I wasn t tired at all. I worked real hard at a high workload right there near the conclusion when I was pulling the rock boxes up. We knew that was going to be hard, plus the fact that we were racing around a little bit towards the end, trying to get everything thrown into boxes and getting all the pieces put together. I expect my heart rate ran up pretty good right there, but I had a lot of energy and reserve at that point, because we had been sort of taking it easy all through the EVA. Everything was, with a few small exceptions, accomplished with a comfortable workload. We didn t have to work hard throughout the whole timeline, and I knew I could afford to race around there for 5 or 10 minutes without jeopardizing the operation at all. They called for a status check and I gave them one and we proceeded, but there wasn t a problem with respect to available energy and reserve.

ALDRIN   I think the fact that you re well cooled off enables you to absorb a fair amount in an increase of activity before it manifests itself. The oxygen flow rate concerned me a little bit preflight because I found, in doing some fairly strenuous exercise in the thermal vacuum chamber, that the first indication I got was that there was not quite enough circulation of air or oxygen to breathe. It tended to get a little stuffy in the helmet. I think all of us who have been through this business know a good bit about the pace of activities following insertion, which is rather leisurely taken. However, you can get wrapped around the axle doing a lot of different things that aren t required - many of them are doing things just to say, Yes - you can add more and more solutions. Therefore, to carry out a minimum rendezvous effort is not, as I would see it, a very tiring task to look forward to after descent and a prolonged EVA. I think we would have been fully capable of carrying out a lift-off and rendezvous.

ARMSTRONG   We handled one.

ALDRIN   You just are not going to get any sleep while you re waiting for it to be completed, but you re certainly not going to be completely bushed chasing yourself around the cockpit. With the automatic radar lift-off and rendezvous are fairly leisurely exercises. I guess I d have more concern about Mike s ability to continue, because he s quite active moving back and forth and doing a lot of manual tasks with the sextant that we didn t have to do.

ARMSTRONG   We cleaned up the cockpit and got things pretty well in shape. This took us a while and we had planned to sleep with our helmets and gloves on for a couple of reasons. One is that it s a lot quieter with your helmet and gloves on, and then we wouldn t have any mental concern about the ECS and so on having two loops working for us there.

ALDRIN   We wouldn t be breathing all that dust.

ARMSTRONG   That was another concern. Our cockpit was so dirty with soot, that we thought the suit loop would be a lot cleaner.

ALDRIN   I guess the question is - Can you keep it cleaner? I guess you could keep it a little cleaner, but there are so many things going in and out that it s almost impossible to avoid getting a significant amount of lunar material in there.

ARMSTRONG   A couple of comments with respect to going to sleep in the LM. One is that it s noisy, and two is that it s illuminated. We had the window shades up and light came through those window shades like crazy.

COLLINS   Why didn t you pull the window shades?

ARMSTRONG   We had them closed. A lot of light comes through the window shades. They re like negatives and a lot of light will shine through.

ALDRIN   You can t see what s going on outside, but you can come quite close to it.

ARMSTRONG   For example, you can see the horizon out there through the window shades. There s that much light that comes through. The next thing is that there are several warning lights that are very bright that can t be dimmed. The next thing is that there are all those radioactive illuminated display switches in there. Third, after I got into my sleep stage and all settled down, I realized that there was something else shining in my eye. It turned out to be that the Earth was shining through the AOT right into my eye. It was just like a light bulb. If I had thought of that ahead of time, we could have put the Sun filter on or something that would have cut the light out. The next problem we had was temperature. We were very comfortable when we completed our activities and were bedded down. Buzz was on the floor and I was on the ascent engine cover. We were reasonably comfortable in terms of temperature. We had the water flowing and the suit loop running. We had to have the suit loop running because our helmets were closed. After a while, I started getting awfully cold, so I reached in front of the fan and turned the water temperature to full up, MAX increase. It still got colder and colder. Finally, Buzz suggested that we disconnect the water, which I did. I still got colder. Then I guess Buzz changed the temperature of the air flow in the suit.

ALDRIN   Yes. We fell victims to a time constant. Once we noticed it going bad there wasn t anything we could do about it. In addition, because we were trying to minimize our activity and stay in some state of drowsiness, we didn t want to get up and start stirring around because it would be that much harder to get back to that same state again. So we tried to minimize our activity. We underestimated how much light was coming in through the windows. There must have been a significant amount of light and heat coming in and just being reflected off the surface. We had no feel for what gas-flow setting we should have had, because we d been on the cooling all the time up to that point while moving around. I m not sure that there s much control over that anyway. We finally disconnected the oxygen flow.

ARMSTRONG   But that requires you take your helmet off, so that you can breathe when you turn the suit disconnects. This means that it gets noisy again, and all you hear is a glycol pump and stuff like that. This was a never-ending battle to obtain just a minimum level of sleeping conditions, and we never did it. Even if we would have, I m not sure I would have gone to sleep.

ALDRIN   I don t know who was on BIOMED at the time, but I feel that I did get a couple of hours of maybe mentally fitful drowsing. I ll have to say that I think that I had the better sleeping place. I found that it was relatively comfortable on the floor, either on my back with feet up against the side or with my knees bent. Also, I could roll over on one side or on the other. I had the two OPS s stacked up at the front of the hatch, so there was ample room on the floor for one. But there wasn t room for two. To cut down on the light level, we re just going to have to do something with the window shades to make them more effective. I think sleeping with the helmet will keep the cooling down and is probably a good reasonable way to go as long as you re going to keep the suit on. Unless some change is made, we d never even think about taking the suits off.

COLLINS   Apollo 12 is planning to take their suits off. With the longer stay time and a couple of EVA s, they re planning to take their suits off.

ALDRIN   I think they ought to think a little more about it. I don t know what the temperature would be in there. I got the impression that it was a lot cooler outside the suit than it would have been inside. I don t feel that having the suit on in 1/6g is that much of a bother. It s fairly comfortable. You have your own little snug sleeping bag, unless you have some pressure point somewhere. Your head in the helmet assumes a very comfortable position. Even out of the helmet, you don t have to worry about what you re leaning against. Your head doesn t weigh that much, and will very comfortably pick just about any position. I just don t see the real need for taking the helmets off.

ARMSTRONG   I didn t mind sleeping on the ascent-engine cover. I didn t find it that bad. I made a hammock out of a waste tether (which I attached to some of the structure hand holds) to hold my feet up in the air and in the middle of the cockpit. This kept my feet up about level with or a little higher than my torso.

ALDRIN   Well, you were back out of the mainstream of the light except for the windows in the AOT. I think we could fix that up and obtain a more horizontal position or the capability to roll from one side to the other. That s just something that has to be worked out. It wasn t satisfactory. If we had known then what we know now, we could have preconditioned the cabin a little bit better. We needed to start at a warmer level by turning the water off, thereby storing a small amount of heat.

ARMSTRONG   That s just one of those areas that didn t occur to us. It clearly needs some more work.

10.53 LEC

ARMSTRONG   The LEC worked as expected; however, I have a few comments worth noting. The primary one is that the LEC was a great attractor of lunar dust. It was impossible to operate the LEC without getting it on the ground some of the time. Whenever it touched the surface, it picked up a lot of the surface powder. As the LEC was operated, that powder was carried back up into the cabin. When the LEC went through the pulley, the lunar dust would shake off, and the part of the LEC that was coming down would rain powder on top of me, the MESA, and the SRC s so that we all looked like chimney sweeps. I was just covered with this powder, primarily as a result of dirt being thrown out by the LEC. This also tended to bind in the pulley. I felt like there was enough silt collecting in the pulley that it was actually binding. Fortunately, Buzz was able to help a great deal. He actually put the majority of the force into pulling the boxes up from the top end, rather than me from the bottom end. I was standing at a very severe angle, which prevented me from using as much force as I had planned for pulling. The ground was too soft and my feet slipped easily. I was leaning over at approximately a 45-degree angle. I had one foot behind me so that if my foot slipped, I wouldn t fall down. The surface was worse. I think the angle and so on were about the same, but I did not have the footing. I couldn t get the footing in this soft powder that you needed to do that job.

ALDRIN   There are several points that tend to make footing more difficult. One is the powdery, graphite-like substance. When it comes in contact with rock, it makes the rock quite slippery. I checked this on a fairly smooth, sloped rock. It was quite easy to get this material on it, and the boot would slip fairly easily. That factor tends to make one more unstable. The second point is that the surface may look the same, but we found that in many areas (with just very small changes in the local surface topography) there would be unexpected differences in the consistency and the softness of this top layer. For example, we might find in some areas where there was just a small slope that when we were on the edge of this slope, there would be little change in the thickness or depth at which we penetrated. In other places, we would find we had put our feet down and we would tend to depress this surface to a new location, as if there were a different depth of the more resistive subsurface. These two factors gave us a low confidence level in our balance and footing setups. To keep the LEC coming smoothly on the inside and to have my pull on it in the appropriate direction so that it neither tangled up near the pulley end nor tended to move or slide the pulley as it went out the hatch, I found that I was completely unable to look out the window at the same time. It was a question of my looking at the LEC, talking to Neil, and hoping we were coordinated. It would be nice to work this over more and try to find some way to maintain visual contact back and forth. I didn t find that easy to do.

11.0 CSM CIRCUMLUNAR OPERATIONS

COLLINS   In general, CSM circumlunar operations went smoothly, and there were no surprises. The spacecraft operated normally; it didn t have any failures.

COLLINS   There wasn t much navigation to be done. I did use P21 several times to pin down the time of arrival at the 150 degree W meridian, which was the prime meridian on the map. It was a simple and easy thing to use P21 to get that information and update the map. The map worked fine with the time tick marks, as long as you are in an orbit of approximately 2 hours time. The map is a useful tool in helping locate where you are with respect to the ground.

COLLINS   The operation of P22 was easy. The procedures that I had condensed into a checklist on the LEB panel were more than adequate. I always went to P22 early, got AUTO optics, and pointed at the landmark far in excess of 50 degrees trunnion. I sat there with a PROGRAM ALARM until such time as the trunnion angle came down below 50 degrees. At this time, I punched off the PROGRAM ALARM and the optics then began to track. I found this was an easy way to operate the system. I had the center couch underneath the left-hand couch for EVA. It was easy to move from the LEB up to the MDC. I found that window two or preferably window three could be used to give you an idea of where you were relative to the landing site. I could look out either of those windows and see all the landmarks approaching. When I got fairly close, all I had to do was leisurely wander down to the LEB, look through the optics, and be ready to mark. The problem was I didn t know where the LM was, and the ground didn t either. There is too much real estate down there within the intended landing zone to scan on one, two, three, or four passes. On each pass, I could do a decent job of scanning one or two grid squares on the expanded map. That map is the 1:100,000 map called LAM 2. The ground was giving me coordinates in the grid square coordinate system that were as much as 10 squares apart. This told me they didn t really have much of a handle at all on where the LM had landed. As I say, it was just too large an area for me to visually scan. I used AUTO optics each time I looked at the area they suggested. I never did see the LM. I don t have any suggestions for future flights. You have to know with considerable accuracy where the LM is before you can mark on it. If you knew where it was that accurately you wouldn t really need P22 to refine your estimate. Perhaps a different Sun angle would yield the possibility of a flash of specular light off the LM skin giving you a clue. I looked for flashes and never saw any.

11.4 MSFN

COLLINS   MSFN worked fine. I was using AUTO on the high-gain antenna. It worked well. The ground was conscientious in updating AOS and LOS times. I don t think that s really necessary. If you re in a near nominal trajectory, as we were, it s an easy thing to do if you have good COMM. If the COMM is intermittent, you can waste 4 or 5 minutes trying to read back and forth AOS and LOS times which really are not required. When the LM is on the surface, the command module should act like a good child and be seen and not heard. The communications with it should take on a negative reporting method.

11.5 PLANE CHANGE

COLLINS   Plane change was not required. The plane change procedure of uplinking a new REFSMMAT and gyrotorquing the platform around to that new REFSMMAT is a tedious procedure. I m not sure that the gyrotorquing is the way to go. A few days before the flight, we abandoned that gyrotorquing method in favor of coarse aligning to the new REFSMMAT. The gyrotorquing took an excessive period of time and had no protection against gimbal lock. We could not even predict in which direction the platform would gyrotorque. That was the story we were given. Some thought should be given to a better procedure for doing that.

11.7 SLEEP ATTITUDE

COLLINS   The procedure was worked out fairly well. I don t recall any mention about deadbands. The ground, in all cases wanted a 10-degree deadband. This was something they asked for in real time. I think it would have saved some chatter over the radio had all this been worked out and put into the flight plan. I needed the control mode and the four or five DSKY operations that are necessary to achieve a 10-degree deadband. Had they been printed in the flight plan, I think that would have helped.

11.8 PHOTOGRAPHY

COLLINS   I thought photography worked out well when I was in there by myself. The amount of time I devoted to photography was somewhat limited by the fact that I was doing P22 each and every pass. P22 was not compatible with good photography. I probably would have spent more time taking pictures had it not been for the question of the LM landing location and the need for the additional P22 s. I did use the intervalometer. I ll have to wait and see how those pictures came out. I feel the command module should carry plenty of film, and I think the key to getting some good pictures from the command module is having the luxury of being able to expose lots of film without worrying about running out of film.

11.10 MONITORING LUNAR ACTIVITY

COLLINS   There was some difficulty with the ground S-band relay. The preflight agreement was that all my transmissions would be relayed to the LM, and all LM transmissions would be relayed to me unless that mode of operation, because of systems failures or other problems, became too cluttered. At this time, the ground was free to amputate that relay mode. In flight, it did not work out that way. The relay was rarely enabled. I gather that this was because there was a ground switching problem. I would have preferred to be receiving continuous S-band relay from the LM, and I felt somewhat cut out of the loop, although it was not a safety problem. I felt out of the loop during the extended periods of time when the relay was not in effect.

11.11 VISUAL MONITORING (MONOCULAR OR SEXTANT)

COLLINS   I did not use the monocular because I did not have the monocular. It went to the surface with the LM. I don t believe it would have been of any use in looking for the LM. The sextant is a more powerful and steadier instrument. It was not possible for me to find the LM on the surface with the sextant.

11.12 CO_2 CANISTER CHANGING

COLLINS CO_2 canister changing was the same as when three people are in the spacecraft.

COLLINS   Maneuvering to support lift-off was worked out well pre flight, and I followed it that way. I couldn t see the LM, but I did nonetheless go through the motions of maintaining the proper attitudes so that my radar transponder would be available in case the LM wanted to lock on. The CSM solo operations were fine. I was at ease about going to sleep and leaving the command module unattended. That didn t bother me at all. I would have guessed preflight that it might have, and it might have if I had had some failures prior to this time.

12.0 LIFT-OFF, RENDEZVOUS, DOCKING

ARMSTRONG   Feed water measurement was performed and the numbers were passed to the ground. I don t remember what they were. First, we zeroed the scale and then with the empty bag on, we took the bag off and reported the RCU weight, with the RCU and not the bag on. Then, we put the water in the bag and reported that weight. That s about a full bag of water.

ALDRIN   Throughout all of this, I didn t have a real high confidence level of the accuracy of what we were doing.

ARMSTRONG   One full bag of feed water is a lot.

ALDRIN   I would think that a volume measurement might be just as accurate.

ARMSTRONG   A volume measurement was the initial plan. That was discarded based on its accuracy.

ALDRIN   The ground had concluded that the water level was lower in my PLSS. It would have seemed to me that that would have been the one to measure, but that wasn t the idea from the beginning. Since they had some indications that consumption was higher on mine, it would have been better to verify that one. We ll see what we get on that. We were given an update on consumables, and we have already talked about the sleep period. They were looking at your BIOMED during the rest periods. As far as I know, we got up just about on schedule and started our activities. It might have been a good bit ahead of schedule, maybe a half an hour or something like that. To try and identify just what our position was, the ground

wanted us to go through a P22 radar track of the command module. We had done this once, maybe twice, in the simulator, and on the surface, it looked like a fairly involved task. Once having run through it in the simulator, it s fairly straightforward. It turned out to be quite a simple operation. Before doing this, we configured circuit breakers and went through a DSKY computer check. I m not sure why it was felt we needed to do this. These were notes as to how we were to handle a P22, option 1, no update. If we got a 503 alarm, we were to key in a proceed and leave the tape meter in altitude/altitude rate so it wouldn t drive into the stops - if it were on range and range rate. It would have been much easier to do a VERB 95 before starting it, because that s evidently what they meant. We went through an LGC-self-test and brought the AGS back on line and then proceeded into the P57. I might point out a few things on the previous day s P57. The yaw left tended to move the one star I wanted to use, Capella, out of the right rear detent. The Sun was in the rear detent and generally obscured it, even though it was not visible in the detent. Its light level was sufficiently high so that no stars could be seen in the rear detent. The Earth was in the forward detent, and due to the yaw left, it was also in the right detent.

12.1 APS LIFT-OFF

ALDRIN   We had another update from the ground instructing us not to go to AGS in the event that the LM engine didn t ignite and not to make a manual start. We agreed that we would wait a REV. Everything worked according to the checklist. We just emphasized that we did use the lunar align mode in the AGS and did not align the AGS to the PGNS, so it lifted off with its own reference system. It did have a PGNS state vector instead of the manual one that we could have given it in the LM slot. Lift-off, or at ignition, we waited until the last 2 or 3 seconds, or almost simultaneously, Neil depressed the abort stage and threw the engine arm switch to ascent and I proceeded on the computer. It might have been a second after the T-zero that any motion was detected. There was, as I recall, an appreciable bang of the PYRO s and a fair amount of debris that was tossed out at the same time that we did detect first motion. It was a fairly smooth onset of lifting force. There wasn t any jolt to it. Yaw started gradually; it was not abrupt either in starting or ending. As a matter of fact, I really didn t notice it. I was looking more at some of the gauges and the altitude rate, both in the PGNS and the AGS. It seemed to take quite a while before we accumulated 40 or 50 feet per second. The pitch maneuver, as seen from inside the cockpit, was not in any way violent or very rapid as we were expecting. We seemed to have a good altitude margin looking down on the surface. It wasn t something that you d describe as a particularly scary maneuver. I felt that we had adequate altitude rate at the time for that type of a maneuver. Right after the pitchover, I could still look out to the side and see the horizon. We could verify out the window what our pitch angle was.

12.4 VELOCITY AND ALTITUDE

ARMSTRONG   Velocity, altitude, altitude rate, and attitudes were consistent with the ascent table that we were monitoring. AGS and PGNS were consistent in attitude as frequent crosschecks on the attitude indicators showed and also in altitude rate, which was being read off the DEDA and compared with the PGNS value of H-dot.

ALDRIN   A couple of years ago, we had a simulation rigged up that tended to give us the sensations in the cockpit that you were liable to experience during LM ascent. We did this in the DCPS and they rotated us back and forth. Based upon this and many ascent simulations in the simulator, watching the rate needles pop back and forth, and the arrow needles wipe back and forth, I expected quite a roller coaster ride of whipping back and forth. Nothing could have been further from the way it actually turned out. It was a very smooth wallowing type of an ascent with far less excursions. Maybe the total rates were approximately the same, but the physical effort of them was not at all objectionable.

ARMSTRONG   The rates and attitude errors and attitude changes were consistent with the simulations. The physiological effect of these was much more akin to the description presented by the Apollo 10 crew of their ascent engine burn. It was very pleasant. It had a Dutch roll mode and relatively low frequency. It was not at all distracting toward your ability to monitor the ascent quantities that were significant. It was a very pleasant and unusual trajectory.

ALDRIN   It was quite easy to look out the window and pick up craters as we approached them. We were keyed to look for the Cat s Paw or anything in the close vicinity to the landing site. I did see several craters, none of which I could positively identify as being the Cat s Paw or in that immediate vicinity. The track looked good as we came up and went by Ritter and approached the crater Schmidt. Communications were excellent throughout the lift-off. We had backup S-band angles at 3 minutes. We didn t need to change any of those. We did accomplish everything in the checklist. The balance couple came off; we were called on the START button at a minute or so after lift-off. Changed the film frame rate at about 3 minutes from 12 frames a second to 6 frames a second. Throughout the remainder of the trajectory, I monitored the targeting quantities in NOUN 76, looked at the countdown time in NOUN 77, then picked up the DELTA-V to go in NOUN 85, and cross-checked it back with VI to compare it with the trajectory. The numbers agreed very closely in H-dot and VI. The altitude looked like it was coming right in on the targeted values, and the AGS agreed quite closely. The V to go, in address 50, did differ a good bit from what I was reading in NOUN 85. However, the AGS gave slightly different targeting. Its targeting is done on a different computation cycle, and I attributed the differences to that. The RCS quantity looked good, and the ascent feed seemed to be operating quite well. To avoid any rush approaching insertion, I elected to open the shutoff valves at about 700 to 600 ft/sec to go. I opened them one at a time, turned off the ascent feed, and closed the cross feed. As we approached 50 ft/sec to go, we still had good pressure in both ascent tanks. Of course, that was one thing we were looking at right up to lift-off to make sure we were feeding on both tanks. I think we inserted with 700 or 800 psi in both helium tanks. Approaching 50 ft/sec to go, we disarmed the engine and it was an AUTO cut off.

ARMSTRONG   I think the over-burn was about 2 ft/sec, and we nulled those.

ALDRIN   There was a certain amount of bounce to them, but since we didn t have anything over 1 or 2 in Z-component, we were able to get the X-component down to near zero, I think 0.1 or 0.2. The out-of-plane residual was small, in the-order of 1, but not over 2. The AGS showed about 8 ft/sec out of plane, and it was, as I mentioned, operating on an independent alignment. VERB 82, as I recall, showed something like a 47-mile apogee. We didn t have the radar to confirm the insertion, but MSFN was quick then to give us a good orbit. The AGS agreed very closely with the PGNS. We got our range rate from the CSM.

ARMSTRONG   It was a satisfactory range rate.

12.5 ATTITUDE

ALDRIN   We got the attitude hold and balance couple on. I don t think we reset abort stage and engine stop immediately. We held off on those, disabled the TTCA s, and designated the radar down out of the field of view in preparation for the alignment. We configured the switches, stopped the camera, and progressed on with aligning the platform.

12.6 PGNS AND AGS

ARMSTRONG   The initial platform alignment planned use of Acrux and Antares as the stars, knowing that Acrux, based on our simulations, would be close to the horizon. I had an alternate in case it was too bright down there. When I AUTO maneuvered to Acrux, it was below the horizon and I couldn t see it, so I chose the first alternate, Atria and Altair. I AUTO ed, so I went out of the program. I re-entered P52, going to star 34, Atria, AUTO maneuvered to the point, and it wasn t in the field of view either. Both of those stars had been in the field of view in all simulations. We terminated the program and reentered at Antares, I think. Is that right?

ALDRIN   37 and 34 are what I have.

ARMSTRONG   We reentered at Nunki, which we knew would be in the field of view. While I was getting marks on Nunki, I had Buzz look up something that might fit with Nunki to be a good second star, and I guess you came up with Atria.

ALDRIN   Yes. It was up in the field of view at that time.

ARMSTRONG   By this time, of course, the stars were rising at a rapid clip, and we could go back to Atria and be quite sure it was in the field of view.

SPEAKER   Which one did you try first?

ALDRIN   Acrux. That wasn t in, and neither was Antares.

ARMSTRONG   Neither was Atria. No. We came back at Atria and got it, and the horizon was in the field of view during the mark. But we had satisfactory marks. We got all zeros on our star angle difference and very small torquing, indicating that our graph, our prelaunch alignment, was quite good.

ALDRIN   I think the largest one was in roll; and, of course, that doesn t affect the insertion as much. The most critical one is the pitch and that had 00064. The one that intrigued me was yaw (which will affect the out of plane insertion), and that had 406. The yaw that we had before that was based solely upon the star alignment that we used before lift-off. It went with the gravity, so it indicated that we had a very good azimuth alignment on the surface. The gravity was certainly adequate to do the job.

12.7 RENDEZVOUS NAVIGATION

ARMSTRONG   It was our intent to pick stars here that would be in the field of view and require a minimum amount of maneuvering and time to get through the alignment and would end up back in plane so that we would be in a place where we could turn the radar on, designate the acquisition, and start getting marks so that we would have a good solution for CSI. Somehow or other, all this planning didn t work out on those stars. Why our simulations did not correctly place those stars relative to the horizon, I don t know. They didn t, so we wasted a little time and a little fuel.

ALDRIN   Even with these problems, we did quite well because we finished about 28 - 27 minutes before CSI and were able to proceed with getting the radar to lock on. That was accomplished without any difficulty. We got one VERB 1 NOUN 49 that we accepted. Before entering the program, we had VERB 95 then loaded to W-matrix. The enable updates - only one of them (the first) failed to pass the test, but it was significantly small, so we proceeded on it. While Neil was doing the alignment, I queried the AGS to see what it thought of the insertion and what it thought the CSI maneuver would be. It came up, just based on the insertion vector, with 15.5 DELTA-M and 51.3 ft/sec.

12.10 ASCENT CAMERA

ALDRIN   The camera was set up with settings as in the checklist, and inserted (pasted) into the checklist at TIG minus 2 was a notation of camera on. At that point, since we were starting at 12 frames per second, it was too early to bring the camera on. I would estimate something on the order of 30 to 40 seconds into the ascent before the camera was turned on. In looking down at the time of the pitchover, I could see radiating out many, many particles of Kapton and pieces of thermal coating from the descent stage. It seemed almost to be going out with a slow-motion type view. It didn t seem to be dropping much in the near vicinity of the LM. I m sure many of them were. They seemed to be going enormous distances from the initial PYRO firing and the ascent engine impinging upon the top of the descent stage.

ARMSTRONG   At the completion of the pitchover, you could easily detect visually that a strong positive outward radial rate had been established. There was no concern about attitude or falling back toward the Moon. I observed one sizable piece of the spacecraft flying along below us for a very long period of time after lift-off. I saw it hit the ground below us somewhere between 1 and 2 minutes into the trajectory.

ALDRIN   It s very difficult to conceive of such lightweight particles like that just taking off without any resistance at all. It s easy to think back and say that they would do that. But it just seems so unnatural for such flimsy particles to keep moving at this constant velocity radially outward in every direction that I could see out the front window. I don t recall seeing any impact with the ground, but there were sizable pieces.

12.13 UPDATES
FOR CSI

ALDRIN   The ground gave us an update of 51.5 ft/sec for CSI with a 1 ft/sec out of plane. I have the values logged down here for what the PGNS came up with, and it eventually settled down on 51.5, also, Mike s solution agreed with the AGS at 51.3, and we elected to burn our solution without any out-of-plane component.

ARMSTRONG   I was just amazed that we had four solutions within 0.2 ft/sec for CSI. That never happened before.

12.15 RCS/CSI
BURN

ALDRIN   I might point out two reasons why we didn t get a backup chart solution. One of them was the alignment. It took a little more time. I think we could have gotten a range rate at 28 and still gotten a good solution; however, the range rate that we were reading at that point was about 51 ft/sec. This was less than the values that were acceptable for the chart. In other words, it exceeded the limits for the rendezvous charts, and since we did end up with a 15-mile DELTA-H and had a good nominal insertion, the only thing I can attribute it to is the command module not being in a circular orbit having enough eccentricity to perturb the R-dot from what it should have been. I think this is another indication of where a late trajectory change was not completely analyzed to see what effects it had. Certainly, we had nominal conditions, but the trajectory change did result in range rate values that exceeded the ability of the chart to cope with them.

SPEAKER   How about the handling?

ALDRIN   The nulling of residuals with the thrusters, even with two jet operations, produced a pronounced difference in translating with just the ascent stage. Each time you hit the thrust controller, the vehicle behaved as if somebody hit it with a sledge hammer, and you just moved. There is no doubt about the fact that the thrusters were firing.

ARMSTRONG   It s a very light, dancing vehicle, and this is true in attitude also. It s very unusual, and the fact that we got five zeros on that alignment, I think, is just a matter of being consistent with all the other good luck we had that day. It certainly was more difficult to do than the upstaged alignment where the vehicle was a lot steadier, and we didn t get results that were that good.

ALDRIN   It was sporty; there s no doubt about it. It appeared that with the automatic tracking and the wide deadband of the radar that it was not bouncing all over the sky. I guess I anticipated that it might have been even sportier than it turned out to be, even though it was a difficult job doing precise aligning with it. I think the 10 mission indicated that. They thought that they had a light weight vehicle, but, of course, they had much more fuel on board than we did.

ARMSTRONG   We did not find as severe a reaction to operating in PGNS AUTO as had been earlier reported. I can t confirm just what their configuration was in terms of the DAP and vehicle inertial, but our combinations made the vehicle fly quite comfortably in PGNS AUTO. We used that mode more or less intermittently with PGNS pulse. We almost did all the manual flying in PGNS pulse, and the remainder of the time, we were in PGNS AUTO. Burns in PGNS attitude holds were generally done with VERB 77.

ALDRIN   That lightweight a vehicle did appear as though it was not an easy task to make either X or Z-axis burns. Of course, all burns were Z-axis burns. To make them, and at the same time avoid having residuals of a fairly sizable number (at least less than 1 ft/sec) is quite difficult. We did end up with minus 0.2, plus 0.7, minus 0.1. The AGS agreed fairly close again, showing the greatest difference in Z, which I think is attributable to the rotation of the burn when loaded into the AGS. The radar stayed locked on throughout the maneuver. We started updating right on schedule, changing the W-matrix to the flight plan values. We checked the numbers just before we changed it. I think we could recall them, if somebody wanted them. 1900, maybe something like 15.3 milliradians. Then we started to work on the plane change. What I did was make use of VERB 90 and ask it what out-of-plane condition it had right now. This saves a little bit of time in not having to load in numbers. We were coming up with things on the order of 1 mile out of plane and 2 or 3 ft/sec. The actual solutions that both vehicles came up with were: our first one was minus 3.2, Mike had minus 2.3; our final one was minus 2.9. As small as they were, we cancelled the plane change maneuver to get more tracking data.

12.18 RCS/CDH BURN

ALDRIN   At CDH, we took out time as computed by the CSI program for the CDH maneuver and voiced in the maneuver to Mike to put in his P76. When you re really getting precise, the question arises what to do with residuals on the order of a couple of tenths. Do you take advantage of them or ignore them? We chose to ignore these small amounts and not thrust. The exception was the out-of-plane condition, and we were handling that as a separate item anyway. DELTA-H varied between CSI and CDH solutions anywhere from 15.3 to 15.7. In general, the CDH maneuver decreased in magnitude. Even on the ones we had in the CSI program, it came up with 19.3 ft/sec and settled down to 18, which I think is indicative of the non-circularity of Mike s orbit. We had no concrete evidence of that really. Our procedures had not called for finding what his orbit was. The ground never did tell us what his orbit was and what we ought to expect for CDH. I think we were kind of left on our own coming up for CDH as to what was an acceptable burn. The data card gives a nominal H-dot of 4 ft/sec. We had 18.

COLLINS   Four is for circular CSM orbit.

ALDRIN   Yes, and that s what you re supposed to have.

COLLINS   I didn t.

12.20 TARGETING PGNS AND AGS

ALDRIN   I had components here for the AGS maneuver: CDH 9.1, 2.4, and 14.6. As per the procedures loaded in the PGNS maneuver, the AGS was updated with the PGNS for CDH.

12.21 UPDATING AGS WITH RR DATA

ALDRIN   After CDH, things seemed to be working so smoothly, and DELTA-H seemed to be so constant that I elected to start putting radar data into the AGS. This seemed to be accomplished without an undue time burden. I ve got the data here that indicates what the AGS solution was for TPI with only AGS update going into it. I think this will be fairly interesting to some people. In other words, it solved for a TPI.

12.23 RCS/TPI

ALDRIN   We burned the PGNS at TPI and then I left the AGS residuals, which are a measure of the difference that it would have solved, and they were on the order of 2-1/2 to 3 ft/sec. Everybody zeroed in on about the same maneuver for the TPI. I guess in the LM you want to delay committing yourself, since you re picking the angle option, to saying exactly what time TPI is going to be until as late as possible. Unfortunately, this presents a burden on the CMP, because he s got the time option. He wants to know what time we re executing it. We gave him a time, and it changed by maybe 30 seconds.

12.27 MIDCOURSE CORRECTIONS

ALDRIN   The first midcourse correction was less than 1 ft/sec. I think we gave the values to the CMP, and he put them in external targeting. The second midcourse correction was about 1-1/2 ft/sec. We burned them in components. I guess it s up to you on angular rates. We picked up range rates from that point on. In a moment of confusion about this time, I observed a significant non-zero lateral deflection in my cross pointer, which I interpreted as being a lateral line-of-sight rate indicating some out-of-plane velocity. This was just a misinterpretation, however. I had to switch in landing radar computer rather than radar line-of-sight rate. So we were actually reading in either AGS or PGNS a version of out-of-plane velocity at that point. I can t explain why that was indicated to be a large number. There wasn t a real number. The line-of-sight rates were, in fact, very low. And as I remember, it was indicating something like 7 ft/sec.

ARMSTRONG   The whole thing from once we finished alignment was just a very leisurely running through of what we had done many times before. Where we were familiar with it was a relatively simple operation. Rendezvous with the PGNS is a piece of cake as long as everything s working. When you start getting PROGRAM ALARMS and radar won t go in, it gets pretty hairy. This happened during several SIM s. You start chasing yourself around the cockpit. But with things working fine, it s simple. It does require close coordination with the other vehicle to keep the flow of information going back and forth. The ground didn t bother us at all. They were watching what was going on, and they called up confirmation of our CDH solution.

12.31 BRAKING GATES

ARMSTRONG   Braking was pretty much on the braking schedule; no problems there. The line-of-sight rates were small and easily controlled. The line-of-sight rate indicator gave us proper indications of line-of-sight rates. The line-of-sight rate indicator does not work like the simulator in several areas. The most significant is when the radar antenna goes from a Sun line-of-site rating back to zero; it does not do it instantaneously as in the simulator. It takes about 5 seconds for the antenna to slow down for a stop for the needle to come to the peg back to zero. Both the sizes of the needle deflections and the rates that they deflect are not correctly simulated in the simulator.

12.36 DOCKING

ARMSTRONG   We stopped braking phase at 50 to 100 feet, insured that both vehicles were in a docking configuration, and at this point, we ran into a problem that we wouldn t have anticipated preflight. Our procedure was for the LM to get into station-keeping position 40 feet out in front of the command module plus X-axis, pitch over 90 degrees so that the X-axes are colinear, then yaw left 60 degrees so that we are in the docking attitude with the command module. It was obvious when we got to this point if we pitched the LM over 90 degrees, we would be looking directly into the Sun. We knew that would be an unsatisfactory lighting condition for docking. So the alternative would be to roll the LM 60 degrees, pitch down, and then you d be in the same attitude and would have prevented the Sun coming into the window. After arriving at that attitude a discussion between the LM and the command module indicated that we weren t quite far enough, so I rolled a little farther, pitched over, and waited looking through the top window. We were asked to rotate a little farther by the command module to line up the docking aids and get the proper alignments. We complied and promptly maneuvered the vehicle directly in the gimbal lock. I wasn t aware of it because I was looking out the top window. No doubt, we were firmly ensconced in gimbal lock. We had all the lights on, the DAP was not operating anymore, we had no control outputs, clearly no CDU outputs were being processed, so we just put it in AGS and completed the docking in AGS.

ALDRIN   And I don t think the AGS is a good system to dock in, or PGNS either.

ARMSTRONG   This was just a goof on our part. We never should have arrived at the conclusion from any series of maneuvers. However, that s how it happened. It wasn t significant in this case, but it certainly is never a desirable thing to do. There s nothing catastrophic about it here, but I m sorry that somehow or other we hadn t studied the docking maneuver a little bit more carefully and recognized that there might be some attitude constraints in the maneuver that we hadn t considered.

ALDRIN   The few times that we d done that previously we ended up approaching docking with the Sun more along the line of sight to the two vehicles. This was more our concern, arriving at the docking point a little bit late. If you arrive there a little late and the line-of-sight motion happens to be such, the Sun is going to be pretty close to where the command module was. In this particular case, it was about 90 degrees away. After getting in that attitude (or getting docked), to have a PGNS operating, I aligned it to zero and went through the quick alignment procedure. I got the PGNS back in operation again and figured it was not a known REFSMMAT. There were no post-docking maneuvers planned by the LM, so to get both systems the same, I then aligned the AGS to the PGNS. Both of them lost their reference, but both of them were 00 and as far away from any future gimbal lock as they could be. That might have been a better way to operate anyway.

COLLINS   The rendezvous procedures from the command module view point were about as well worked out, I thought, as they could be with the existing command module computer structure and with the degree of participation necessary by the CMP. I have always felt, and I still feel, that the system is designed in such a fashion that the CMP is too busy during the rendezvous procedure. Although I was able to keep up with the timeline quite well, I felt that I was devoting too large a percentage of my time to the job and that I really was poorly placed to cope with any systems problems or any other difficulties or abnormalities that might have come up. I don t propose any sweeping changes from mainline Apollo. It would be fruitless to do so, but I really think that for future vehicles the rendezvous should be something that is relatively straightforward, something which does not require literally hundreds of simulator hours to master the procedural aspects of. I think, as we get into these lunar-exploration flights, the crew is going to be forced to devote more and more of their attention to what they re going to do once they ve arrived, not just to working out the procedures for how to arrive. I really think that. From the command module viewpoint, with one man inside the command module, I think the procedure should be simplified, and if that requires a greater degree of automation, then I think we ought to have more automation. I had a solo book which combined features of various other publications, the idea being I wouldn t have to chase around the cockpit; I would have everything under one cover. This concept worked well. I recommend it highly. The only funny I had during the rendezvous was the VHF ranging kept breaking lock. I had a good lock-on during ascent just as I had during the simulations. I was surprised when, after insertion, VHF ranging broke lock. I did reacquire, but from then on, the thing broke lock 25 times during the course of the rendezvous. Sometimes, I could immediately reacquire with the reset switch on panel 9. Other times, it was not possible to reacquire. I would have to go VERB 88 ENTER to lock the VHF ranging data out of the computer, and then at some later interval, I would get a good reacquisition and do VERB 87 ENTER to allow that data to come back into the computer. It was possible for me to tell, after a little practice, whether I was going to get a good lock or not by listening to the tone during the lockup procedure. There are three tones, two of which are in the audible range. If it was going to be a good solid lock, the tones would be very clear and sharp just as they are in the simulator. If it was not going to be a good lock, if the lock was going to be unsuccessful, the tone sounded very scratchy and had a lot of static. After the third tone had completed its cycle, the numbers would appear very briefly on the EMS and then they would almost immediately go to zero, indicating the thing had broken lock. I used a technique of setting the mission timer in the lower equipment bay to the nominal LM lift-off time of 124 hours, 23 minutes and 25 seconds. At the instant the LM lifted off, I started that clock running. I kept two times; the LEB time was flight plan time. If I followed it, I could with a high degree of accuracy tell you where I should be in my procedures book. I left the MDC clock undisturbed, so that all the updates and communications with the ground could be done in true and correct time. It worked well for me. I practiced it in the SIM s. I was influenced by the fact that the digital event timers had a poor history of reliability and that the digital event timers in spacecraft 107 had been replaced once and further had little funnies in them during tests. If you trust the digital event timers to count down to burns, then probably my procedure is an unduly complicated one. On the other hand, it is workable. I found it an aid in running through this, despite the fact that we were a couple of minutes off nominal. I have some numbers on breaking lock. I first got lock-on during the latter part of the ascent burn. It broke lock at 124 hours and 31 minutes, reacquired immediately, and broke the second time at 124:34. It broke twice thereafter in rapid succession. I relayed my out-of-plane solutions to the LM. They were, after insertion, on the order of 1 ft/sec, and the LM Y-dot minus 1.0. I had my own Y-dot plus 1.4. CSI solutions compared fairly close with the LM and the ground solutions. I think Buzz has reported those numbers previously. I was in an orbit of 63.2 by 56.8, which could explain some of the up-down component in the CDH solution, as well as some R-dot peculiarity the LM experienced. I had some eccentricity in your orbit after CSI. I had you 49.5 by 46.1. I think the combination of those two screwy orbits could explain lots of R-dot dispersions and up-down components.

ALDRIN   I asked the computer what time we were going to get to apogee, and it was only a couple of minutes off to CSI time.

ARMSTRONG   Yes, but you didn t know where his apolune was.

ALDRIN   That s right. Ninety degrees away.

COLLINS   I don t think it s worth our spending a lot of time here trying to hash out these numbers. I just mentioned them for the record. A plane change was not required during the burn. I still think that it s possible under some disperse circumstances to have a large plane change required following an ascent from the lunar surface. This plane change might have to be done by the command module using the SPS. This is something that had never had procedures worked out for it. I did invent a procedure. I don t think FOD liked it although they didn t have any better procedure. I would suggest that the FCOD come up with a procedure that MIT and FOD and the Center agree might be used to allow the command module to whip around and make an out-of-plane SPS burn. Now, the one that I invented was sort of sneaky. It took P52, the platform align program, and told P52 to align the platform to a landing site which arbitrarily was said to be 35 degrees north latitude. Of course, this is completely phony but it cocked the platform off 35 degrees in roll, so that when you yaw out of plane either left or right, you ll go above the cherry or below the cherry, because the platform has been rolled out of the way. It worked well in the simulator. I don t know what the objection to it is. I had a little procedure drawn up, and it was included in the rendezvous book. If this is not a good procedure, then it should have been substituted for something better. You need to have in your hip pocket some quick way of whipping that command module around 90 degrees and burning SPS.

ALDRIN   You didn t want to do it by just taking it 45 and asking for a good preferred alignment?

COLLINS   But you have to get a new REFSMMAT from the ground and everything else.

ALDRIN   Another way, you load the burn in P30.

COLLINS   I see what you re saying.

ALDRIN   Then go into P40, find out what it is, set the REFSMMAT flag, and then go into it.

COLLINS   Yes. That s another way of doing it. You can do it that way as well. This P52 way was just quick and simple and dirty. Another little funny I noticed (maybe it s something that I overlooked in my training) was after CSI when I went to P20. P20 would not track the LM; in fiddling around, I found that if I recycle the optics zero switch it would track the LM. Now, as a matter of practice, I had always left the LEB in optics manual and optics zero to zero. The reason you leave it that way is because of failure modes in the CDU s, which are shared with the thrust vector control if you re going to burn the SPS. In the simulator, when you get back down into the LEB, all you have to do is take that optics zero switch and throw it off, and P20 will immediately start tracking the LM.

ALDRIN   Did you have it in CMC?

COLLINS   Yes, CMC. But on two occasions it wouldn t do it. I found if I cycled the optics zero switch it would track the LM. I don t understand this. It s as if there is a funny in that switch in regard to the optics power. When you first turn optics power on for P52, you have to cycle that switch back to zero for the program to be aware that it has been zero. Otherwise, you get into CDU difficulties. It s something similar to that. Anyhow, after CSI when I went to P20, the sextant would not track the LM until I had recycled the optics zero switch. How I knew to do that, I don t know. It was just trial and error. VHF ranging broke lock again along about plane-change time. It broke lock again at 126 hours. We ve already discussed the CDH solutions. I had hoped to get some sextant marks immediately after CDH, but prior to going into P34 this was questionable because of the position of the Sun. I thought that I could probably get three or four marks before Sun shafting prevented it. I was wrong. I couldn t get any marks at all. After CDH, I was able to get VHF marks only for a little while. In the meantime, I went into P34 and I had a very slow COMP cycle the first time through on P34. Up until this time, the AUTO optics had been doing a smooth job of tracking the LM. I noticed that this smoothness disappeared a few minutes prior to the TPI. It became quite jerky. I made a little note here - the LM tracking jerky in sextant, and DAP excessive pitch thruster firings. It seemed like there was a little flurry of pitch thrusters firings along about this point as well, which I don t have any explanation for. Everything progressed normally through TPI. It was along about midcourse time when I first saw the LM coming up from below. It looked like the doggone LM was riding on rails. There was absolutely no line-of-sight rate that I could see. It really looked great to see the LM coming up from the surface. For the first time, I had the feeling that that son of a gun was really going to get there in one piece. Midcourses were very small. Braking was done entirely by the LM. I was completely passive, and that s all I have to say about the rendezvous. Docking we did in CMC, AUTO, narrow deadband under DAP control. Neil made the crude alignments to get the correct side of the LM pointed toward the COAS. Then I made the final adjustments. I estimated that I contacted the LM just about exactly dead center and at a slow but adequate closing velocity. I would guess slightly in excess of 0.1 ft/sec. Despite this fact, I couldn t tell the instant of contact. The empty ascent stage is light enough relative to the command module that when the two vehicles touch, it s just sort of like pushing into a piece of paper. The LM recoiled enough that they could feel it in the LM, but I couldn t really feel it in the command module. I thought I was getting there, and I thought I was getting there, and I finally was fairly sure I had contact. I looked up for the third or fourth time, and I did have two barber-poles indicating that the capture latches had made. At this time, I looked out the window, and the situation appeared static. I threw the switch from AUTO to FREE, so that I was in CMC, FREE. I looked out the window again - this was all going pretty fast now - I would say this was 3 seconds after contact. The situation looked like it had previously, that is, the two vehicles were statically joined together with no motion. At that time, I fired the bottle. No sooner did the bottle fire than a yaw gyration started between the two vehicles. I m not sure whether it was a result of the retract cycle beginning or whether it was a result of the LM firing thrusters toward me. At that time, this static situation became very dynamic, and a fairly large yaw excursion took place. I would say that relative to the LM I rapidly went to about a 15-degree yaw right angle. I put the CMC, FREE switch back to CMC, AUTO. This enabled the hand controller in rate command and minimum deadband. I made manual inputs to yaw back over towards the centerline, and there were a couple of other oscillations en-route. I can remember thinking, I don t think we re going to get a successful hard dock this time. I m probably going to have to let the LM go and try again. About that time, the docking latches fired, and we were hard docked. I would guess that the time interval from firing the bottle to hard dock was about 6 to 8 seconds. This is probably a pretty normal retract time. Things were happening fairly rapidly, and the oscillations had built up almost exactly at the time I fired the bottle which was primary 2.

ARMSTRONG   I can add a few comments here from the other side. At the time we felt the contact - which really was difficult to feel - it was a very low bump sound, or touch in the tunnel we fired plus X RCS in the LM as per the preflight plan. Shortly thereafter, we also observed significant attitude oscillation. I guess it would be primarily right roll as observed in the LM. We were in AGS RATE COMMAND minimum deadband and, in addition, plus X. As soon as the attitude deviation started, I left the plus X off and called for Buzz to give me MAX deadband in the thrusters so we wouldn t be firing a lot of attitude thrusters. Then I took control and manually maneuvered the vehicle back toward colinear status. About that time, it snapped us in there and locked the latches.

COLLINS   I didn t like the idea of these two vehicles being joined together just by these two little capture latches. I was in the habit of firing the bottle the first time it appeared; the two vehicles had been joined together and the situation was static. I never gave these oscillations a fair chance to develop. Maybe a better thing to do is delay firing the bottle until you are sure the oscillations are not going to develop. Although it was sort of alarming there for a second or two, this way did work and it was within the envelope. I m not sure if I had it to do over again that I would do differently. It depends on what caused the oscillations to get started. It could be the thruster firing of the LM or it could be some other cause. If it s the thruster firing of the LM, then you ought to delete the thruster firing on the LM. I m not really sure you need that thruster firing on the LM.

ALDRIN   I m not either.

COLLINS   If it s some other cause, then the thruster firing of the LM is probably not a bad thing.

ALDRIN   It should tend to give some stabilizing effect to the LM. You d like to have some control system that s holding the LM fairly close to where you want it to be. I think automatic is probably able to catch sooner than manual. Because you re looking up this way, it s pretty darn hard to maintain a close position. That argument says that you ought to be in some kind of automatic rate command system.

ARMSTRONG   I think we have to admit that this was one area, in retrospect, that we gave less thought to than it probably deserves. During simulations, none of our simulators is able to duplicate this kind of dynamics. We saw some film that had been taken of a McDonnell study. We saw these and observed what their recommendations were. That s what was incorporated in our docking plan. That really was devised to get the capture latches in. I really suspect that everything we experienced happened after the capture latches were engaged. The results of that study really weren t pertinent to this particular phenomenon. We hadn t experienced any trouble at all on your previous docking. That was just as smooth as glass.

ALDRIN   It seems to me that it s not too good a mode to be working in. You re tempted, if the thing starts to move on you, to touch the stick. As soon as you do that, you have now reset a new attitude that may not be what the combined systems are going to be happy with; and if it s not, it s going to fire.

ARMSTRONG   That s right. I m not sure that a lot of thought on our part in this area would have made the situation any better.

COLLINS   No. That s right.

ALDRIN   I don t think we got a tremendous amount of guidance out of the AOH or anybody. It seemed to be, however you want to do it. You can do it this way or that way. They are both acceptable means in the AOH. It seemed to me there were two ways to be acceptable, and this was with primary guidance control. We didn t have primary guidance control because of the gimbal lock problem. It seemed to me that the book treated that subject a little lightly. Wasn t it written for LM active?

ARMSTRONG   Yes.

COLLINS   We gave the subject very little training time, but had we given it a lot of training time, I m not sure we could have come to any different conclusions.

ARMSTRONG   It did bite us a little bit.

ALDRIN   It s worthy of concern because if you do prang something the consequences are time consuming and nasty to have to go through.

ARMSTRONG   This one got to us and, for one reason or another, we didn t understand it well enough. I suggest that the next crew spend a little more time than we did in this area and try to improve on the procedures.

ALDRIN   All other dockings were done in PGNS.

COLLINS   This was the same procedure from the command module. The only difference was that the LM ascent stage was considerably lighter.

ARMSTRONG   The LM control configuration was different.

COLLINS   Yes, I meant from the dynamics of the command module viewpoint. I had the feeling that going to FREE under these circumstances was a mistake.

ALDRIN   You don t have a good choice of deadbands. Half a degree seems to me to be too tight for this operation, and 5 degrees is much too loose.

COLLINS   Flag it as a problem. I don t have a solution.

12.38 POST DOCKING CHECKS AND PRESSURIZATION

COLLINS   When I went into the tunnel this time, I had that same strong odor of burnt material. Again, I checked everything very closely and couldn t find anything wrong. All the decals and checklists were well worked out for the probe and drogue. I was glad to see it work. I never had much confidence that our tunnel was going to work as advertised, but it sure did. I was very happy to see the tunnel, the probe, the drogue, and all that stuff part company and go along with the LM.

12.39 TUNNEL OPERATIONS

COLLINS   We went through an extra operation, and this is something that we never practiced jointly. It was my intent to take the probe out, the drogue out and put those two items inside the command module. I guess it was your intent to take them out from your side and put them inside the LM. I just happened to beat you to it. It really wasn t very efficient the way I did it.

ALDRIN   I thought you were going to do it.

ARMSTRONG   I had it in my mind that I was going to do it.

COLLINS   The flight plan didn t mention it. It sort of implied that the guys were going to do it, because it said to remove and stow the tunnel hatch, and then it said to notify LM crew they could open their hatch. It didn t mention the probe and drogue. When I came to that, I thought, they just left that out of the flight plan. I said, Stand by one. Then I got the probe and drogue out and stowed them onboard in the command module. This was an extra operation because subsequently they had to be transferred to the LM. This is another area where we couldn t say that we had smooth coordination. I knew how to do my end, and the LM knew how to do their end; but we hadn t sat down and discussed who was going to do precisely what.

<u>12.40 TRANSFER OF LM EQUIPMENT AND FILM</u>

ARMSTRONG   The equipment transfer and cleaning back contamination procedures were done essentially in the manner that was planned. We had a couple of small differences. We decided we wanted to bring the LEVA bags, and the LEVA s, and the EVA gloves back with us for post-flight examination. We brought the whole ISA, interim stowage assembly, with all its transfer gear into the command module. The intent was to unload that, restow it in the command module, and then take the ISA back into the LM. We didn t do that. We brought the ISA back in the command module with us. That s a 1-pound item or something. We were able to get through that procedure about on the planned timeline.

ALDRIN   As a matter of fact, they were thinking about moving up TEI.

ARMSTRONG   Well, as it turned out, our LM jettison time could not have been moved forward a REV.

ALDRIN   Because of attitude.

ARMSTRONG   We couldn t have made it really.

ALDRIN   Because of the attitude?

ARMSTRONG   No, we just couldn t have gotten through in time.

COLLINS   We were an hour, maybe an hour and a half, ahead of time.

<u>12.41 VACUUMING EQUIPMENT</u>

ARMSTRONG   I was concerned that it might take us a lot of time to clean the LM, and I was also concerned that we would have a lot of free-floating lunar dust in the cockpit going back to insertion. We really wondered at engine cut-off whether we wouldn t be completely engulfed in soot and be unable to take our helmets off for the alignments. However, there wasn t much dust, and we couldn t figure that out because

ALDRIN   The stuff seemed to stick to things and stay there.

ARMSTRONG   I thought we d tramped a lot of it with us, but it didn t bother us.

ALDRIN   I wiped it up with my suit on the floor.

ARMSTRONG   We did clean with the vacuum cleaner as best we could. That vacuum cleaner has a very low suction, and more time was required than we planned to do the cleaning job. We were afraid it wouldn t be done to the degree of completeness that we had hoped for. We were able to clean the suits satisfactorily with a scrubbing motion. However, there wasn t a large amount of free contaminate in the LM. We wore the suits back into the command module and restowed them in the L-shaped bag after a drying-out period. The LCG s were also stowed with the suits in the L-shaped bag. The suits were relatively clean, but they had a lot of residual smudges on them.

ALDRIN   There was no hope of getting that off.

12.43 STOWAGE OF SRC s

ALDRIN   The bags for the rock box, I think, could have some better labeling on them. You want the box to be mounted correctly in the command module so that one g or the g forces of entry will push the material down towards the bottom of the box instead of the top. But nothing really tells you how you put the box inside the bag. You can put it either way. We learned by the way the lettering was, you had to put the bag on the box upside down to the way you normally think. It would help if the zipper went around the bottom instead of around the top; so I think that some more labeling would be in order just to make sure that no one puts the box in the bag upside down. I don t know how critical that is, but it s worth noting.

ARMSTRONG   Stowage was planned, plus we had a large temporary stowage bag completely filled with command module trash, food wrappers, and so on, which was transferred to the LM to clean up the command module volume.

13.0 LUNAR MODULE JETTISON THROUGH TEI

13.1 LM JETTISON

ARMSTRONG   LM jettison went as planned.

ALDRIN   Was there ever any intentions to track the LM after jettison.

COLLINS   No. That was never even discussed.

ALDRIN   I don t understand why we left it in VHF ranging mode and left the track light on.

COLLINS   I have no idea. We never had a DTO on it, or to my knowledge, it was never even discussed.

ARMSTRONG   The separation was slow and majestic; we were able to follow it visually for a long time.

COLLINS   The LM held its attitude extremely well. I don t know what mode you left it in, but I thought when the explosive charge fired, it would sort of start going ass over tea kettle. It must have been in some good attitude hold mode, wasn t it?

ARMSTRONG   We could watch the jets fire to hold attitude as it went away.

ALDRIN   It was in MAX deadband, AGS ATT hold. It seemed to me that, right at the time of separation, as the LM moved away, I could see some cracks that had developed in the outer thin skin of the top part of the LM in the gray material that forms an area around the docking cone. However, according to the ground it held pressure. I couldn t see any other damage that had been caused by blowing the tunnel.

COLLINS   The only comment that I had is that the separation burn was something that MPAD had changed their minds about a time or two. Originally, it was going to be 1 ft/sec horizontal retrograde. Then for some reason, they wanted it 45 degrees up from horizontal, and they wanted 1 ft/sec retrograde component or a total burn of 1.4 ft/sec. I don t have any preference one way or the other. It just seems like that s a fairly simple thing, and they ought to get their desire worked out early in the game and not have that be a late, last minute change, because it just makes for last minute conversations on unimportant things.

ALDRIN We didn t put the helmets in the LEVA s did we?

ARMSTRONG No.

ALDRIN Looking back on it, I think it would have eased the stowage problem in the command module.

ARMSTRONG Yes, but there was a reason for that, and that s that the LEVA s and the EVA gloves were both awfully smudgy. The choice there was to leave them sealed up in the LEVA bags rather than to get that soot out into the command module.

COLLINS The activities prior to TEI were leisurely. The updates were passed up in good time, we passed our sextant star check. In general, the usual sequence of P30 and P40 is one that has been well worked out and TEI had no surprises up until TIG time.

COLLINS Well, we took a few photographs prior to TEI, but essentially we spent the time preparing for the burn. We didn t do any television prior to TEI.

COLLINS Those criteria were ones that had been hammered out for a long time. We didn t have any argument with them. Essentially it was a 2-second over-burn, if confirmed by EMS reading of minus 40 ft/sec. We came close to shutting the burn down manually - I ll get into that a little bit later.

COLLINS At TIG, this was the first burn with CSM only. I had my rate needles on 5/1 and I did that because I think it s a good mode to be in if you re worried about any sort of abnormal dynamics. They re much more readily apparent on the sensitive scale.

COLLINS At TIG, I noticed more rate-needle activity than I had seen in previous burns. I had a start transient of probably 0.4-ft/sec activity on the rate needles in both pitch and yaw; there was very little attitude deviation. It was just a fairly rapid oscillation of both the gimbal position indicators and the rate needles and it damped itself down I d say within the first 10 or 15 seconds of the burn. In roll, the vehicle was deadbanding. Instead of plus or minus 5 degrees, it appeared on my attitude indicator to be more like plus or minus 8 degree roll deadband and it was banging against the roll stops fairly crisply. It would cruise over, hit deadband and jets would fire, and it would go back the other way.

This roll deadbanding was quite obvious during this burn as opposed to the other burns. I think all these indications are normal. They were just somewhat exaggerated during the first 20 seconds of the burn compared to the more damped case of having the LM attached. The EMS counter moves out pretty swiftly and it was difficult for me to estimate exactly when I might have minus 40 on the counter. The Isp of the engine must have decreased or something; at any rate, the burn duration was longer than predicted and when burn time plus 2 seconds had elapsed, I had thought that I would have minus 40 on the EMS counter by the time I could get the thing shut down. There was some doubt in my mind as to whether it was shutting itself down automatically or not; so, at burn time plus 2 seconds and some small fraction, I turned both EMS DELTA-V - or both DELTA-V - normal switches off. I think just a fraction of a second prior to this we got a good automatic shutdown. At any rate, our residuals were very small; so either we got a good automatic shutdown followed immediately by my turning the switches off or else I shut the thing down manually and was just extremely lucky in that it coincided with the PGNS residuals. For some reason, that burn duration was a little bit longer than I would have expected. LOI, you remember was shorter than we had predicted and this was the next burn to follow LOI, so I was sort of surprised that it did take longer than normal.

ALDRIN   The PUGS was a little bit unpredictable based upon performance during LOI. The fact that I couldn t catch up with the increase and it got ahead by about 0.4 or 0.5, something like that, plus the preflight briefing that that would be the case was why I left the switch in INCREASE. We lit off and got through the initial guidance and I looked at the meter and it was showing down in DECREASE, which struck me as not being what it should do. I expected it to be in INCREASE, but I thought Well, maybe this is a characteristic such that early in the burn it does this sort of thing. So I left the switch where it was to try to catch up. I guess in the meantime that the two numbers - where one had been bigger than the other - had changed positions, in addition to the fact that when it says INCREASE, you throw it in the INCREASE direction. It s not at all obvious during a burn if one is a little bigger than the other. You re not sure whether the needle is believable or not, so I left it in INCREASE and it seemed as though it was getting farther apart and the needle was staying down; so contrary to what we had been led to believe, I put the thing down to DECREASE just to see what was going to happen. Sure enough, it stopped the divergence of the two numbers. We didn t have a long enough burn for it to get right to zero, but it was within 0.2. Anyway, it was a little different than what we had expected. I guess, if you really want to play that game, you might need to write some cues or something on there so you don t misinterpret anything. It worked out well. But it was unusual and that might have something to do with burn time.

ARMSTRONG   We tried something different on this flight. The ground computed a postburn state vector, a predicted postburn state vector and put it in the LM slot. After the end of the burn, we could call up VERB 83 and get an R and R-dot from our state vector over to the predicted state vector. It came out real close, 0.7 mile and 0.8 ft/sec, indicating (it s kind of another double check) that we really did get the burn that we thought we were going to get. That s not really any kind of requirement if everything works. It is a nice kind of thing if you have an SPS problem or if you take over with the SCS in the middle of the burn when your computer is working okay, but the guidance isn t working. You can use that vector in your hip pocket to find out how good of a switchover you did and how close your SCS burn came out.

14.0 TRANSEARTH COAST

14.1 SYSTEMS VERIFICATION FOR COAST

COLLINS   All the systems were GO; there wasn t anything to do.

14.2 NAVIGATION, NAVIGATIONAL SIGHTINGS, AND OPTICS

COLLINS   We didn t do any onboard navigation. Our flight plan called for doing it only in the event of COMM failures. The optics worked normally on the way home.

EVAPORATORS:
ACTIVATION &
DEACTIVATION

COLLINS   We did not activate either the primary or the secondary evaporators until just prior to entry; so, during trans-earth coast, those were not in the system.

14.4 PASSIVE THERMAL CONTROL

COLLINS   Passive thermal control three modes - we didn t have three modes, we just had the one mode. We always rolled G&N control at 0.3 deg/sec; that procedure we ve already talked about. There were no differences in trans-earth, although the geometry of the vehicles was a lot different and I thought that the command module by itself would go unstable more quickly. Neil thought it would not, and he was right. It was very stable on the way back, just as it was on the way out.

ALDRIN   The LMP would have preferred pointing north. However, there was an added advantage in that we got to look at the Magellanic clouds by PTC-ing at 270.

ARMSTRONG   To look at the earth, to look north, you had to get upside down.

COLLINS   Yes, we went out in 090 pitch angle and came back 270 pitch angle. It s macht nichts to me; I don t care one way or the other.

14.5 EXCESSIVE MOISTURE ON TUNNEL HATCH AREA

COLLINS   There was a little tiny bit of moisture up in there at various times. On the way home, there was less than there had been earlier. The last time I checked was at 180 hours or thereabouts.

ALDRIN   You thought it was less? I don t remember much moisture at all.

ARMSTRONG   I thought it was more on the way home.

ALDRIN   I did too. We made use of the ECS hoses.

COLLINS   Yes, I put the hoses up there and there s one comment in here. Here it is - 180 hours, dry as a bone.

ALDRIN   That was after we put the hoses up there.

COLLINS   Prior to that, there was a little bit of moisture up there and I did wipe it off with a towel sometime after TEI.

ARMSTRONG   I could go into the tunnel usually and wipe my finger around the hatch up there and come back with a wet finger.

COLLINS   Well, you could see little beads of moisture like on a beer bottle or something like that.

COLLINS   There weren t great globs of moisture and, as I say, at 180 hours, it was dry as a bone. When we came to entry, we wiped excessive moisture from the tunnel hatch area. That leads me to believe that it has something to do with the routing of those hoses. If you really cram a set of hoses up in that tunnel as far as it will go and sort of wedge the hoses up around the side of the hatch as far as you can, it might help keep the circulation pattern up. That would keep it fairly dry.

ALDRIN   We shot up a batch of film right after TEI. We pitched down and picked up a good attitude to photograph the moon out the hatch window.

COLLINS   Yes, we took a whole lot of what I think should be real good pictures.

ARMSTRONG   We made a lot of color-comparison checks.

ALDRIN   Well, we haven t mentioned anything yet about the color as viewed particularly and I guess it is one thing people are going to be listening or looking for before they debrief us. I think that it makes some difference which window you re looking out because the windows do seem to have a little bit of a coating on them. I got the distinct impression that it depended on how you looked out of a particular window, what angle you looked out of it, to tell you just what color you were going to see on the surface. It didn t look the same out of each window. That could answer a lot of questions about the differences that people see and I m sure that not every spacecraft has the same coatings on the windows. I don t know how significant it is though.

14.6 FUEL CELL PURGING

COLLINS   Fuel cell purging was normal on the way back.

14.7 CONSUMABLES

COLLINS   We finally - we almost caught the RCS budget. Last hack on that, we were 1 percent down and on the hydrogen and on the oxygen we were very close to nominal. Whoever figured those out did a good job.

14.8 SPS MIDCOURSE CORRECTION

COLLINS   None was required on the way back. We did have one midcourse of 4.8 ft/sec which we did with the RCS.

14.9 MIDCOURSE LUNAR LANDMARKS

COLLINS   That s not applicable.

14.10 STAR/EARTH HORIZONS

COLLINS   That s not applicable.

14.11 ECS REDUNDANCY

COLLINS   We did not investigate any of the redundant systems of the ECS.

14.12 DAP LOADS

COLLINS   DAP loads were as called out in the flight plan; I don t have any comments on those. We widened up the DAP deadband PTC to 30 degrees, which is really sort of a waste of time in that DAP PTC procedure, because as soon as you widen the deadband, you turn all 12 or 16 of your RCS thrust switches off. It really doesn t matter whether the deadband is wide or narrow, the thing is incapable of firing any thrusters anyhow. The DAP loads as written in the flight plan were satisfactory.

14.13 IMU REALIGNMENT

COLLINS   IMU realign was all right. Throughout the flight, I was able to get satisfactory IMU alignments during the PTC at 0.3 deg/sec. This is a fairly fast rate, and it feels uncomfortable. You have to go to RESOLVE MEDIUM, and you have the feeling that you are lucky to click the stars that pass through the center of the reticle pattern. It s not really possible to track smoothly and hold the star in the center and make a very precise mark. However, the star-angle differences came out usually 00001, so I guess : that the accuracy is well within the limits that you would call satisfactory.

14.14 COMMUNICATIONS

COLLINS   Again, the ground was changing between OMNI B and D in PTC. When we were stopping PTC, we were getting little snatches of the high gain. Difficulties with that system were traced mostly to ground-switching problems, although you would have to say it is a fairly cumbersome system using the four OMNI s and the high gain. I don t have any suggestions for improving the operating procedures.

ALDRIN   It would be nice if the ground had control of that OMNI switch to select any of the four.

COLLINS   Yes, that s true. Right now, the ground can either switch between high gain and D Dog or between D Dog and whatever is selected on the switch to its left, which is normally B Baker.

14.15 BATTERY VENTING

COLLINS   Battery venting and waste dumps were all normal, just as they were on the way out.

14.16 POWERING UP & DOWN OF SPACECRAFT

COLLINS   We only powered a few items down each night. We really maintained power for the entire flight, and that was a mode of operation I enjoyed, not having to power down.

14.17 TELEVISION

COLLINS   We made a goof on our last television show. We left the circuit breaker out, which allows the monitor to be operable without transmitting. Consequently, we lost a lot of the entry data. It s the one on 225 called S-band, FM transmitter, data stowage equipment flight bus. Of course, the entry checklist didn t mention checking that circuit breaker, because the people who wrote the entry checklist had no idea that it would be out because of a television program hours prior. I guess the TV checklist doesn t mention it either as best I can recall.

ALDRIN   I was sort of disappointed in the ground not catching that. It seemed to me that they might want to make some checkout of the tape because they had control of it before entry, or because we called out to them that the talkback barber-pole didn t go gray.

COLLINS   We did lose control of the tape because the circuit breaker was out. I believe that we and the ground both got tricked into thinking it was because we hadn t gone to COMMAND RESET. But didn t you tell them once that you had gone to COMMAND RESET and you still didn t have tape control?

ALDRIN   Yes.

COLLINS   To make a long story short, we did inadvertently leave that TV circuit breaker out, and therefore, the taped entry data were lost. They ll still have a lot of information through the downlink.

14.18 COAST PARAMETERS - ANOMALIES

COLLINS   The machine held together beautifully on the way home. I don t know of any anomalies.

14.19 HIGH GAIN ANTENNA TRACKING

COLLINS   High gain antenna tracking was as it always was.

14.20 S-BAND PERFORMANCE

COLLINS   S-band performance was good.

14.21 NECESSITY OF ADDITIONAL IMU REALIGNMENTS

COLLINS   The IMU, by this time, had had its compensation terms updated once or twice, and it was in good shape. I don t recall the longest period of time we went without an IMU alignment, but it was on the order of 12 hours. At the end of this period of time, the stars were still well within the sextant field of view.

14.22 MCC UPDATE

COLLINS   Midcourse correction update was well handled. We only had an RCS burn.

14.23 W-MATRIX

COLLINS   We didn t fool with it; we left it alone.

14.25 PRESLEEP & POSTSLEEP CHECKLISTS

ALDRIN   We talked once about looking into some modifications of the COMM so that you didn t have the two options available, plus referring to another checklist with exceptions. I think there s some way to simplify that.

14.26 PHOTOGRAPHY

COLLINS   We took lots of pictures on the way home, using up the remainder of the film. We took photos of the exterior of the Earth and the Moon at various settings. We ll just have to wait and see how they came out.

14.27 PASTIME ACTIVITIES

COLLINS   What did we do with our free time? We mostly just waited. We had plenty of time to eat, had plenty of time to get rested up. We used simultaneous sleep periods on the way home. Our inclination during preflight was to use staggered sleep periods on the way home. I m not sure in retrospect which is the best way to go.

ALDRIN   I didn t see anything wrong with the way we did it.

COLLINS   I didn t see anything wrong with what we did, because nothing broke. Had we had things start breaking, I m not sure we wouldn t have been better with the staggered sleep periods.

14.30 TIMELINES AND FLIGHT PLAN UPDATES

COLLINS   There was none that I recall.

14.31 MANEUVERING TO ENTRY ATTITUDE

COLLINS   Maneuvering to entry attitude was done easily and early.

14.32 BORESIGHT AND SEXTANT STAR CHECKS

COLLINS   We did not have a boresight star, but the sextant star check passed as it always did.

14.33 ELS LOGIC AND STAR CHECKS

COLLINS   The ELS logic check was done early with the ground looking over our shoulder, and it gave us a GO for PYRO ARM.

14.34 EMS

COLLINS   We checked the EMS out insofar as we could the day prior to entry. I think this is a good idea because if there are any funnies in it, then the ground has a good 24 hours or more to have meetings and decide whether or not all or portions of the EMS are GO or NO GO for entry. The DELTA-V counter worked normally in EMS. Accelerometer bias - I don t really recall that we checked that pre-entry. We just ran through all the self-test patterns, and one of those checks accelerometers when it counts down to zero plus or minus something.

14.35 ENTRY CORRIDOR CHECK

COLLINS   The ground kept reporting our gamma, which was indicating a little steep, 65 something. Then we got closer and closer to nominal as we got closer in, and I don t recall, what our actual gamma was. I think it was 652.

ARMSTRONG   No. 648 was the last we hit.

COLLINS   648 is as close to nominal as you can get.

14.36 FINAL STOWAGE

COLLINS   We had a couple of items, mostly helmets, that did not go according to the entry-stowage plan. The helmets were supposed to go in the food boxes. Only one helmet fit in the food box and that left us with two helmet bags plus two LEVA bags. These four little packages we bundled up and put inside the right-hand sleep restraint and latched down with tie-down cord. That system worked fine. Our first inclination was to put all those bags inside the hatch bag underneath the left-hand couch. However, the ground objected to that because they thought that the bag wasn t stressed sufficiently for that weight during entry, but I think you could have put 10 helmet bags inside the hatch bag and it would have been perfectly safe. That hatch bag is very strong and it s a very convenient place to stow things even of helmet weight during entry.

ALDRIN   We ought to find out what limits North American places on that for entry.

COLLINS   You could grab that hatch bag and pull on it with all your might and you weren t about to pull that thing loose.

14.37 SYSTEMS VERIFICATION

COLLINS   The systems worked fine.

14.38 FINAL ENTRY PREPARATIONS

COLLINS   Final entry preparations were done early with a good checklist.

14.39 CM RCS PREHEAT

COLLINS   CM RCS preheat was not required.

14.40 MANEUVERING TO ENTRY ATTITUDE

COLLINS   We used the system of manually tracking the horizon and cross-checking gimbal angles and horizon positions in the window versus time out from 400,000 feet. The ground had given us several check points at EI minus 30 minutes and EI minus 17 minutes. In addition, we had a little graph that showed for any instant in time what the pitch gimbal angle should be to keep the horizon on the 31.7-degree line on the window. All these checks reinforced our belief that we did have a good platform and that we had a good trajectory.

14.42 CM/SM SEPARATION

COLLINS   CM/SM SEP went normally. The water boiler was in operation during this period of time, which gave the spacecraft a left yaw. I was in MINIMUM IMPULSE a good percentage of this time, and thus it was quite noticeable. I yawed out 45 degrees left, jettisoned the service module, and yawed back in plane by yawing right. When I got a yaw rate started, the water boiler would fight me, the rate would reduce to near zero, and I would then have to make another input. Having gotten back to zero yaw after jettisoning the service module, I noticed there appeared to be something wrong with the yaw-left thruster at this time. It had worked normally for a little while, but after several minutes of operation, it did not. That was command module RCS thruster 16, yaw left. It appeared to be functioning improperly using the automatic coils. When you yawed left, it made some noise, but it did not give the proper response. It would work properly if you d move the hand controller all the way over to the hard stops and use the direct coil. At this late stage of the game, I didn t want to devote any time to troubleshooting or talking about it. I probably should have brought the number 2 system on the line in that axis, but I didn t; and everything else seemed to be working normally. I m just flagging that as a possible systems problem; somebody should look at that thruster and its associated wiring after the flight and see if there s anything wrong with it.

FCOD REP   Did you see the service module?

COLLINS   Yes. It flew by us.

ALDRIN   It flew by to the right and a little above us, straight ahead. It was spinning up. It was first visible in window number 4, then later in window number 2, really spinning.

<u>14.44 0.05g EMS AND CORRIDOR CHECK</u>

ALDRIN   What was the comparison of when the final g light came on?

COLLINS   Twenty-eight seconds, I think.

ARMSTRONG   When the DSKY indication of the accelerometer acceleration read 5, the 0.05g light came on. At that point, the clock read 28 seconds.

COLLINS   The spacecraft was briefly out of the sunlight at 400 K, and all of a sudden the thing lit up and I thought we were starting to get ionization, but it really wasn t that; it was a brief period of sunshine.

ARMSTRONG   I wasn t looking out, but there was a weird illumination. I also thought it was just ionization at the time.

COLLINS   We got the 0.05g light, and I got the 0.05g switch and the EMS roll switch on. We were cross-checking the clock, and this was 28 seconds after 400,000 feet. I did not notice the corridor verification light, either the upper one or the lower one. Both of them could have been on. I was busy at this time checking other things, such as were we holding the right bank angles with the lift vector up and did the g on the EMS agree with the g meter. I was also listening to what Neil was saying about the computer. Of course, our intent was to hold the lift vector up unless we had some considerably off-nominal entry with no communications; so we started to do that regardless of what the corridor verification light said.

15.0 ENTRY

<u>15.1 ENTRY PARAMETERS</u>

COLLINS   Normally we re targeted for 1108 miles from pointer K to the ship. Initially the weather in that area looked good, but as we got in closer, Houston started making grumbling noises about the weather in that recovery area. Finally they said there were thunderstorms there and they were going at 1500 miles. I wasn t very happy with that fact because the great majority of our practice and simulator work and everything else had been done on a 1187 target point. The few times we fooled around with long-range targets, the computer s performance and the ground s parameters seemed to be in disagreement. Specifically, there s an exit velocity and exit-drag-level check that s got to be within certain bounds, and it rarely, if ever, was within those bounds. So, when they said 1500 miles, both Neil and I thought, Oh God, we re going to end up having a big argument about whether the computer is GO or NO GO for a 1500-mile entry. Plus 1500 miles is not nearly as compatible - it doesn t look quite the same on the EMS trace. If you had to take over, you d be hard-pressed to come anywhere near the ship. For these reasons, I wasn t too happy about going 1500 miles, but I cannot quarrel with the decision. The system is built that way and, if the weather is bad in the recovery area, I think it s probably advantageous to go 1500 miles than to come down through a thunderstorm.

<u>15.2 COMMUNICATIONS BLACKOUT</u>

COLLINS   I never paid the slightest bit of attention to that. They read up all the numbers; it s simple that, if you re in a blackout, you can t communicate; if you re either side of the blackout, you can. I guess the ground uses it a little more than that, it can give them more of a hack on where you are relative to the nominal trajectory.

COLLINS   Along about .05g, we started to get all these colors past the windows; Buzz took some movies, which we looked at last night. They don t really show what the human eye sees. Around the edge of the plasma sheath, there are all varieties of colors - lavenders, lightish bluish greens, little touches of violet, and great variations mostly of blues and greens. The central core has variations on a orange-yellow theme. It s sort of a combination of all the colors of the rainbow really. The central part looks like you would imagine a burning material might look. Orangeish, yellowish, whitish, and then completely surrounded by almost a rainbow of colors.

ALDRIN   I thought there was a surprisingly small amount of material coming off.

COLLINS   That s right; there didn t seem to be any chunks as there were on Gemini.

ALDRIN   That s right; there didn t seem to be any droplets or anything coming off. There was a small number of sparks going by; you could definitely see the flow pattern. Looking out the side window, you could get a very good indication of the angle of attack by the direction of motion of the particles. That didn t seem to change too much. When a thruster would fire, you could pick it up immediately, because it deflected the ion stream behind you. I am not sure whether that was because of a roll or whether it was actually changing the direction of the lift vector.

COLLINS   I didn t hear any unusual sound at all during that time.

ALDRIN   No, it seemed to be rather quiet.

COLLINS   Yes, there wasn t any sizzling, popping, or any noises that you commonly associate with entry heating.

ALDRIN   I thought the g constant was quite smooth.

COLLINS   I thought it was smooth also.

ALDRIN   More rapid from a physical standpoint then I had anticipated.

COLLINS   I thought it was slower than the centrifuge. I think it s probably exactly the same time duration.

ALDRIN   Well, I didn t have a meter to look at.

COLLINS   You re more keyed up and time seems to go more slowly. Anytime that I go from zero g to positive g, I get a feeling of transverse acceleration instead of feeling, like that of what it truly is. The first few seconds I get the sensation of body rotation, mostly in pitch. Usually I think we re pitching up.

ARMSTRONG   Yes, I would agree that I felt a little bit of a rotational sensation during the initial g pulse, but it s not disorienting.

COLLINS   We gave spacecraft control over to the computer after we passed all our pitch attitude cross-checks. We gave it to the computer shortly before 400,000 feet. I don t recall exactly when, but a matter of seconds before 400,000 feet. We stayed in CMC, AUTO for the rest of the entry. The computer did its usual brilliant job at steering. We just sort of peered over its shoulder and I made sure that the spacecraft was responding to the bank angles that the computer commanded, and that those bank angles made sense in light of what we saw on the EMS and through other bits and pieces of information. The computer did not fly the EMS the same way I would have flown the EMS. As soon as it got sub-circular, it seemed to store up a lot more excess energy than I thought was reasonable. It was holding on to an approximate 250 miles downrange error. When the downrange distance to go was, say, 500 miles, it would have about 750 miles available at that particular g level we were seeing at the time. I thought this was probably a little excessive, but it hung on until very, very late in the game and then it decided all of a sudden to dump it. It sort of rolled over on its back and gave us a second peak pulse of 6g s getting rid of that excess energy. After that everything was all cross-ranges and down-ranges, and everything made sense. It was essentially on zero error for the remainder of the run. Our first peak pulse was 6.5 as nearly as you can read that thing, and the second one was 6.0. The EMS trace looked more like a roller coaster than a horizontal line. It really climbed for altitude after the initial pulse and hung way up there high. All of a sudden, it decided to dump it, and rolled over on its back and we came screaming back in. That is really a pretty gross exaggeration, but that was the trend.

15.8 DROGUE CHUTE DEPLOYMENT

ALDRIN   I could see the ring departing just a fraction of a second before I felt a small pulse. There wasn t much of a rotation as the drogue chutes deployed. They seemed to oscillate around a good bit, but did not transmit much of this oscillation to the spacecraft. The space craft seemed to stay on a pretty steady course.

15.9 MAIN CHUTE DEPLOYMENT

ALDRIN   The main chute deployment again gave us a small jolt, but not one that would move you around in the seat appreciably or cause any concern. I can t say that I noticed the difference between first deployment and dereefing. It seemed to be one continuous operation.

COLLINS   It seemed to me there was quite a bit of delay before they dereefed. All three chutes were stable and all dereefed and they kept staying that way until I was just about the point where I was getting worried about whether they were ever going to dereef; then they did.

15.10 COMMUNICATIONS

COLLINS   As soon as we got out of blackout, we heard Recovery 1 and Hawaii Rescue 1. Houston, as per agreement, stayed off the air and we pretty much stayed off the air except to speak when spoken to and to let the recovery people know that we were in good shape and that there was no hurry about their recovery operations.

15.11 ECS

COLLINS   We did not have suits on. We brought the primary water boiler on the line as per the checklist; the same for the secondary. They were brought on roughly 45 minutes and 15 minutes, respectively, before separation, something like that. I don t think the secondary boiler really had a chance to do any boiling; however, I believe the primary did.

ALDRIN   At any rate, it started perking away. You can t tell how effective it is by looking at the gauges.

COLLINS   The cooling was very good. Even during the entry itself, we were perfectly comfortable. We didn t have to freeze ourselves out by cold-soaking prior to entry. We didn t go through any cold-soak procedure. It was pleasantly cool throughout the entry, and it was quite comfortable on the water, as opposed to our Gulf-egress training. I think you d get an entirely different viewpoint of that recovery operation with the BIG s, if you started out hot with stored body heat.

16.0 LANDING AND RECOVERY

COLLINS   I felt a solid jolt. It was a lot harder than I expected.

ALDRIN   It pitched me forward with a little bit of sideways rotation. I was standing by with my fingers quite close to the circuit breaker. The checklist fell, and the pen or pencil, whatever I had, dropped. It didn t seem as though there was any way of keeping your fingers on the circuit breakers.

ARMSTRONG   When you are 18 knots away, it looks pretty promising.

COLLINS   I think those procedures for the main chute are well worked out. I think it is 50/50 whether or not you are going to Stable II.

COLLINS   The postlanding checklist worked well. The big item for us was that we not contaminate the world by leaving the postlanding vent open. We had that underlined and circled in our procedures to close that vent valve prior to popping the circuit breakers on panel 250. I d like to say for the following crews that they pay attention to that in their training. If you cut the power on panel 250 before you get the vent valve closed, in theory, the whole world gets contaminated, and everybody is mad at you.

ALDRIN   I have a couple of things noted in the checklist. I don t think any of the flights have ever used the CM RCS preheat. If you miss a circuit breaker, it is not real obvious that you are going to come back and see that circuit breaker later. You do, but it is tucked away. For example, when you get ready to preheat, you push some circuit breakers in and turn the heaters on. You wait awhile then pull some circuit breakers out again. The way the checklist is written, some of those circuit breakers stay in and you wonder whether you ought to go through the mechanics of checking all those things off. The other one is the CM RCS activation. When we got to the point of bringing the various logic switches on, the sequence arm circuit breakers were out. Mike called it to my attention that unless we pushed those in, we weren t going to get any RCS pressurization. We didn t go back and research this at that particular time. I believe that if the checklist people check, they will find that those circuit breakers should be called out to be pushed in at that point. During the CM RCS check, it says to go to spacecraft control SCS, but it doesn t tell you what mode to be in for the check. I think you want a minimum impulse. I think that it is logical that it be called out in the checklist.

COLLINS   It was definitely humid inside. We got about a quart of water in through the snorkel valve. It was definitely humid, but it was comfortable.

COLLINS   Communications were good after we became stable I. Of course, we could not hear anybody in Stable II, because the antennas were in the water.

COLLINS   Battery power was more than adequate for the brief duration we were in the spacecraft. I don t recall the voltage, or you mentioning it.

ALDRIN   On the main chutes after we dumped propellant during the purge cycle, you could see flame coming out of the thrusters and going by the side windows. When we opened up the valves, there was a fairly strong odor of propellants. It didn t last particularly long. It seemed to me we had plenty of time, and it might be advisable to delay that a little bit longer.

ALDRIN   The visibility out the side window coming down was quite good and I felt that you could look out and almost see impact by looking out to the side. This would involve some risk to your neck at the time. I think you could determine levels of 50 feet or less and then put your head back on the couch. I didn t see any need to do that, but the capability does exist. This business about hitting the water without putting the chutes out because of altimeter failure is kind of a Mickey Mouse simulator pad.

COLLINS   I think it is a good thing though. I think the more answers you can mess up inside the simulator, the better it is. They ought to trick you into coming in with your PYRO circuit breakers out, with your ELS circuit breakers out, with your PYRO s not armed, with your ELS logic off, or with your ELS AUTO switch in MANUAL. Any one of those things can really foul you up. To get a successful entry, you have to have the ELS circuit breakers in, the PYRO s armed, ELS AUTO on, and the ELS LOGIC on. Those are important things in the 101 checklist items. Most of them are really not critical, but those few items are. I managed to foul each one of them up at least once during the various simulations. I was glad that I had because I was darn sure going to make positive that each of those switches were in the proper position.

ARMSTRONG   I agree with that. What Buzz is saying is that this lack of information about how high you are is not real. If you are in a lighted condition, and we were in a relatively well-lighted condition during chute deployment, this information is a lot more readily available in flight than it is in a simulator. You can see the clouds coming up, and you are watching yourself go through cloud layers, and then you can see the water down below you. You have a lot of cues as to how high you are which aren t available in the simulator.

ALDRIN   That s true.

16.10
SEASICKNESS

COLLINS   Nobody got sick. We each took a pill prior to entry and a second pill on the water. Those pills are called - Hyacynth and Dexedrine, and they seem to work fine.

ALDRIN   No side effects at all.

16.11 INTERNAL
TEMPERATURE
CHANGES

COLLINS   There didn t appear to be any. We were comfortable on the water, and I guess at the time it probably warmed up a little. We weren t in it long enough to really feel any sudden changes. There weren t any.

16.12 STABLE I OR STABLE II - UPRIGHTING PROCEDURES

COLLINS   We were in Stable II. The float bags worked fine. We were in Stable II 4 or 5 minutes.

ALDRIN   It didn t seem like it was anywhere near as long as it was during the tank or Gulf training exercises.

COLLINS   I am sure the reason was that we were bobbing around fairly well.

ALDRIN   As soon as they became almost full, the wave action tipped it back over.

16.14 INITIAL SITTING OR STANDING

COLLINS   That s right.

COLLINS   I don t think any of the three of us had any of those symptoms.

16.15 INTERNAL PRESSURE

COLLINS   Internal pressure was fine. We used the dump valve as per the checklist, and it worked out well.

16.16 RECOVERY OPERATIONS

COLLINS   Recovery Operations went very smoothly. The swimmer threw the BIG s into us. We put the BIG s on inside the spacecraft. We put them on in the lower equipment bay. Neil did first, then I did after him. Buzz put his on in the right-hand seat. We went out; Neil first, then me, and then Buzz. It s necessary, at least the way we had practiced it, for us to help one another in sealing the BIG s around the head to make sure the zipper was fully closed.

16.19 EGRESS

COLLINS   As we crossed the threshold of the hatch, we inflated our water wings and jumped into the raft. The BIG s swimmer had trouble getting the hatch closed. I don t know why. Neil went back to help him, and he still had trouble. I went back to help, and when I got there the hatch gear box was on neutral and the hatch handle was on neutral. He should have been able to close it. The hatch handle, instead of being up at its detent, was flopping free. All I did was take it and cram it up into the detent. Then he was able to close the hatch. He was really cranking on it. With neutral on those two pawl settings, there should be no impediment to closing the hatch. Even if the hatch handle is flopping loose, there isn t anything inside which mechanically would interfere with it. We finally helped him get the hatch closed. We sprayed one another down inside the raft. There was some confusion on the chemical agents. There were two bottles of chemical agents. One of them was Betadyne, which is a soap-sudsy iodine solution, and the other one was Sodium Hypochlorite, a clear chemical-spray. During our simulations, we used Betadyne in both bottles. They found that the Betadyne broke down the waterproofing in the suit. They made a last-minute change and used Betadyne for scrubbing down the spacecraft, but they used Sodium Hypochlorite for scrubbing us down. I had read about this and knew that there was a change. While the swimmer was scrubbing the spacecraft, I grabbed the other bottle and started scrubbing Neil down. The swimmer got excited and didn t want me to do that. I found out later it was because if you inhale enough of this Sodium Hypochlorite through your intake valve you can cause problems inside the BIG. I m not sure whether you get nauseated, you can t see or your eyes water. You have to be careful and not spray too vigorously around the intake valve. You have to spray your glove and wipe it on rather than spray it directly on. I am sure future recoveries will have this worked out during their Gulf egress training. This is just another example where changes made between the training and the real thing have the potential of biting us.

ALDRIN   I thought the BIG was a well-designed garment. I was rather disappointed in the visibility. When we had our training exercise in the Gulf, I didn t notice as much fogging over on the inside of the visors as I did on the actual recovery. I thought for a while it was on the outside. I dipped down in the water, but couldn t seem to clear it at all. I don t know where it came from. It didn t seem to me that I was perspiring that much on the inside.

ARMSTRONG   I was just about to comment on the same thing. If there were any disadvantages to the BIG, as they were used in this operation, it was the lack of visibility due to condensation on the inside of the visor. It was so bad as to be nearly opaque.

COLLINS   I didn t notice that it was any worse than the Gulf. I could see the helicopters clearly, the sling being lowered, and the swimmers. I could make out enough detail, for example, to read the face of a wrist watch. I could see fairly well.

ALDRIN   You could, but you would have to move it around to a clear spot.

COLLINS   Maybe that s true.

ARMSTRONG   It may have had something to do with the seal between the face and the mask.

ALDRIN   Yes.

ARMSTRONG   How tight that seal was determined whether or not that condensation was excessive or not. Perhaps you had a tighter seal. I think that my seal was fairly loose.

COLLINS   So was mine. You remember, you wanted to tighten my mask.

ALDRIN   I tightened mine down. Mine was pretty tight so that I wasn t breathing in and out of the suit. Maybe that fact contributed to mine fogging up.

COLLINS   Could be. I don t know

ARMSTRONG   I had a loose fitting mask, too. I had the same problem.

16.21 CREW PICKUP

COLLINS   We got into the raft, did our decontamination bit, and they picked us up. The helicopter pilot was real good. You put one hand or foot anywhere near that basket, though, and they start pulling. They don t wait for you to get in and get all comfortable before they retract. Just like a fisherman, they felt a nibble on the end of that line, and he started cranking. Aboard the helicopter, we started storing heat. For the first time I became uncomfortably warm during the helicopter ride. That helicopter ride was as short as we are going to have them during this kind of operation. We debriefed the recovery people out on the ship and told them the same thing. When you get the crew on the helicopter, everybody shouldn t sit back and breathe a sigh of relief and think that the operation is all over; they should keep right on moving. This is the time when the crew is really starting to get uncomfortable. If the crew has to stay in that helicopter 15 or 20 minutes longer than we did, I guess the hood on the BIG would come off. That s a pretty wild guess.

ARMSTRONG   I agree.

ALDRIN   I agree.

ARMSTRONG   I think we were approaching the limit of how long you could expect people to stay in that garment.

COLLINS   It was all right in the raft.

ALDRIN   The roughness of the water didn t bother me too much. The fact that we were getting just a few waves every now and then cooled you off. There was no way of measuring what the inside temperature of the chopper was except that we just started accumulating heat inside the suit.

17.0 GEOLOGY AND EXPERIMENTS

COLLINS   I thought that the maps were more than adequate - those that were carried onboard the command module. The grid system could be improved on. I think the ground sort of, in real time, came to the same conclusion we had - to call each grid the letter defining its lower boundary and the number defining its left-hand boundary and using sort of a Vernier scale across the grid square. In other words, if you want to define a spot in grid square E9, you consider E9 the one whose lower left-hand corner is the intersection of E and 9. If you want to get specific, you say E.9 and 9.13 or something like that; and that defines within that 1 square more specific coordinates.

ALDRIN   The numbers, if yours were the same as ours, had some pattern to them; but they didn t have as much pattern as I think could have been employed. In other words, they could have just gone straight across left to right in each row.

COLLINS   Here s mine. I m talking about the LAM-2 map, and it was okay. It s no jewel of a map, but it was certainly adequate.

ARMSTRONG   This section is going to be difficult to do without the pictures to describe. We re going to take an hour to talk about some of these things that you could talk about in 30 seconds with the picture.

[EDITOR S NOTE] The remainder of the items listed in Section 17 were covered in considerable detail in the air-to-ground transcription and/or Section 10 (Lunar Surface) of this document.

18.0 COMMAND MODULE SYSTEMS OPERATION

18.1 GUIDANCE AND NAVIGATION

COLLINS   I have no comment about the ISS. Optical subsystem B2, light transmittance telescope and sextant - as we said previously, the sextant was a very useful instrument as long as the platform was kept in alignment within plus or minus 0.9 degree. Then stars would be visible in the sextant and it was very useful. The telescope, on the other hand, was a very poor instrument because of the light loss through it, not being able to detect star patterns without a considerable period of dark adaptation. That s all I got on that.

18.1.3 Computer Subsystems

COLLINS   We raised the possibility of making a PTC program. The computer probably has enough memory to do that if you start deleting things like stable orbit rendezvous, stable orbit midcourse P30, a P39 perhaps. We had no restarts or any funnies in the computer.

18.1.4 G&N Controls and Displays

COLLINS   G&N controls and displays were all without surprises. No comments.

ARMSTRONG   I have one comment on the EMS. I think a review of the residuals from each burn that was made would indicate that there is something we don t understand about properly computing DELTA-V_C, because there seems to be a definite similarity between the residuals. It does not seem to be proportional to the burn size or burn time or any of those things. I always end up with 4 or 5 ft/sec of DELTA-V_C.

COLLINS   The other peculiarity of the EMS was during transposition and docking. The EMS functioned normally during the separation from the S-IVB and the subsequent acceleration, but after the turn around, after the 180 degree pitch and the 60 degree yaw, the numbers in the EMS did not make sense. Instead of being around 101 or 100.6 to 101, they were down below 100. Then, in fact, I docked with the EMS reading 99.1, which is completely non-sensible. I don t understand how or why the EMS got jolted off its correct values during that turn around.

18.4.1 SM RCS

COLLINS   On the SM RCS system, we had one quad that was considerably noisier than the others and I don t understand exactly why that was. I don t even remember which one it was. I think it was quad A.

18.4.2 CM RCS

COLLINS   I think there is something wrong with the AUTO coil functioning on thruster 16.

18.5 ELECTRICAL POWER SYSTEM

ARMSTRONG   Any comments?

ALDRIN   Worked like a dream. The initial battery charging was a little surprising in that the voltage was quite high when the battery charger was first turned on. It was up around 39.2 or 39.3. It later went back on down and the amps went up. I don t really understand what the cause of that was.

I called it to the attention of the ground and they seem to think it was normal, but I don t understand it. I got in the habit of using the battery bus indicator whenever I turned the main bus ties on and off, just as a confirmation of doing that. I just mention that for any use of follow-on crews.

18.6 ENVIRONMENTAL CONTROL SYSTEM

COLLINS   There was one funny in the ECS, and that had to do with the primary glycol evaporator outlet temperature getting lower than normal one time during lunar orbit when I was in the command module by myself. It seemed to be a transient condition. The system recovered and began functioning normally. It gave the appearance of the bypass valve having malfunctioned and putting fluid that should have been bypassed around the radiators and through the radiators, resulting in a RAD OUT temp that was too low and a glycol EVAP temp that was too low. And after just sitting there for a while watching it, the system slowly recovered and for the remainder of the flight, the primary glycol loop worked perfectly normally. So there was apparently some transient there which I am unable to explain.

ALDRIN   On the ECS, it seems to me in that rapid REPRESS package we would have a better gauge than that one that goes up to 1200 but has marks every 300 psi. That s not a big thing. You don t refer to it very often, but it s just not a very easy one to read, and it s not very sophisticated.

ARMSTRONG   You might mention this inadvertent operation of the press-to-test valve a couple times.

COLLINS   On the oxygen panel, the emergency cabin pressure regulator push-to-test button - we hit it with our toes several times, and it made the ground nervous to see a sudden inflow of oxygen to the cabin on the TM. I don t know where they pick off their TM. We heard a little hissing noise that didn t pass more than a second or two. For some reason, our feet banged into that area during the first day or two. A couple or three times we did push-to-test that little button.

ARMSTRONG   Any trouble with CO_2?

COLLINS   No, I had no trouble. They wanted a time in and a time out recorded on the CO_2 canisters. I did write the times on the side of the canister, both in felt-tip pen and with a regular mechanical pencil. It would be easy to put some kind of sticker or to provide some place on the side of the canisters to write on. They are very dark metal, and they are very slippery. It s very difficult to write on, even with the felt-tip or pencil so that they will be legible. I recorded in and out times on each of the canisters, but I m not at all sure they will be able to read that information.

ALDRIN   It s kind of silly to record those in and out times on the ones that you jettison with the LM.

ARMSTRONG   On the other hand, it s not much trouble to write the times down when you re making the change.

COLLINS   Yes, if we never got into the LM for some reason, I suppose they would want all that information.

18.6.2 Cabin
Atmosphere

ALDRIN   I noticed occasionally when my eyes would water, there would be a certain noticeable burning. This occurred when water would drip around, and part of that I guess is due to zero g. It s primarily when you wake up in the morning. My eyes would just start to water, and I would notice the burning.

COLLINS   Well, something else I noticed in the way of eye irritation was that the male Velcro that s mounted on the spacecraft would come apart in little tiny bits and pieces. That material would get on your skin. A couple of times I noticed eye irritation in the inner part of my eye, and I d get my finger and peel off a little segment of that Velcro. That happened more than once. That stuff floats around the cabin and can get into your eyes.

18.6.3 Water
Supply System

COLLINS   The chlorine injection port became more and more difficult to use. The chlorine seemed to corrode the metal, and the chlorine injector assembly became covered with sort of a black slushy-looking deposit. I think it was a chemical interaction between the chlorine and/or the buffer and the metal. The friction in the system got higher and higher, and toward the end of the 8 days, it was very difficult to screw the ampoule assembly into the injector.

COLLINS   The filters I think are a good idea, but they need some more engineering done on them. The basic problem is the back-pressure characteristic or the range of back pressures which will result in satisfactory filter operations. The back pressures should be held to a minimum, and, of course, as long as you are just squirting water through the gun into your mouth, for example, the filter seemed to work pretty well. It still allowed some gas to get through. I think under all circumstances some gas got through. I couldn t really measure that, because you can t see the water in your mouth. However, I just had the feeling when squirting the water gun in my mouth through the filter that the water still had some gas in it. But it was a lot better than it would have been without the filter. Hooking a food bag up to the filter or a drink bag changed this situation. It depended upon the individual characteristics of each bag just how much the operation was degraded. Some bags had very nice, smooth openings in them, and some were crinkled and wrinkled.

COLLINS   You really couldn t open up a sufficient orifice behind the valve so that the water gun would be pumping against the back pressure. This, I think, degraded its efficiency. The dispenser in the LEB is a 1-ounce dispenser and without the filter attached, every time you depress the plunger you get a very forceful ejection of 1 ounce of water with a very definite beginning and end to it. With the filter attached, you depress the plunger and you get about a half ounce of water rapidly ejected, followed by a very slow oozing out of another half ounce. In other words, the filter acts as sort of an accumulator for the system. Since that second ounce appears very slowly over a long period of time and hangs there as a globule on the end of the filter, it sort of makes for a leakage problem. Whenever you try to fill a water bag or food bag with either hot or cold water from that spigot down there, you have to wait an awfully long time after the last squirt to let the water come through the filter. Even so, you are going to get a lot of leakage after you disconnect the water bag, because at the instant you disconnect the bag, the back-pressure characteristics are changed and the oozing increases. So, in general, it was just sort of a sloppy operation trying to use the filter with a lot of spillage. You had to really get down there with the towel every time you wanted to fill up the water bag. On a couple of occasions, we really put the back pressure to it. I remember one time, the entry port to one food bag was totally blocked and we were trying to squirt water into a deadheaded system. Under these circumstances, my preflight briefing indicated that relatively irreversible damage would be done to the filter, and we would have to take it up out of the line and go through a drying procedure of several hours duration. We found this was really not the case. We could see the membrane deteriorate, and then little beads would appear on it. Yet if we just let the filter alone for several hours, it appeared to us that the efficiency was restored.

ALDRIN   It seemed to be a mechanical problem, also, of attaching the bags. Without the filter, the nozzle of the gun would stick inside the bag valve, opening with enough friction and with the O-ring fitting tight enough so that you didn t have to push with any appreciable force to retain the bag on the end of the gun. You could confidently squeeze the trigger and squirt the water into the bag without fear of squirting the bag off the end of the gun. But that wasn t true with the filter on there.

ALDRIN   First, the filter didn t have a good locking device to keep it on the gun. When I put the end of the filter into the bag, I found that I had to continually push the bag to retain it on the gun.

COLLINS   That O-ring is just insufficient.

ALDRIN   The opening on the end of the filter isn t quite long enough, either. And I found that to work it better I d have to cut a good bit closer to the nozzle end than the line would indicate. This would let you get the end of the filter farther into the bag without the bag interfering with where the nozzle of the filter wasn t long enough.

COLLINS   We can get together with Al Tucker and show him exactly.

ARMSTRONG   The filter took some gas out, but not all. Of course, efficiency of the filter varied widely depending on what the back pressure situation was.

ALDRIN   The problem with the gas in the bag is one of difficulty in mixing the water and whatever is in there. There is some discomfort when you swallow a fair amount of gas, but the biggest thing, I guess, is the fact that you just pass more gas. Of course, that s a big odor problem in the spacecraft.

COLLINS   I beg your pardon.

ALDRIN   I beg yours.

ARMSTRONG   Let s go on to water glycol system.

COLLINS   Water glycol; no comments. There was the one funny on glycol evaporator outlet temperature in lunar orbit being too low. The secondary glycol loop check when we got a small decrease in the accumulator quantity is something that perhaps should be explained a little better preflight so that it came as no surprise.

COLLINS   Nothing.

ARMSTRONG   Just the O_2 flow sensor in the suit circuit.

COLLINS   Well, that s not really a circuit. That s on the 100 psi line, but I think that s well documented. All we can say is the transducer was sick, or somehow the O_2 flow sensor onboard reading was out of calibration. It read lower than the flow rates we were actually getting, and its degree of accuracy seemed to change with time.

COLLINS   The flow rate again; that s the only thing.

ALDRIN   Yes, I think there is some question as to the capacity of that waste management container and the ability to fully utilize it. I wasn t convinced that we didn t have a good bit more room than we thought we did, but there wasn t any real way of knowing how much additional volume you had available. And I wasn t too successful in being able to put my arm in there and push things down.

COLLINS   I had a smaller arm. I could get my arm down not quite around the corner. I could get it far enough down to sort of try to keep things moving to the bottom of the barrel. Again, I guess I didn t have a precise handle on how much that would hold. I knew that it was just a fairly large compartment. I think we could have stuffed more things in there than we did. On the other hand, we did have a priority system of what we thought should go in there. The smellier it was, the more desirable it was to have it inside that compartment. I thought it worked out fairly well.

ALDRIN   Yes, but I think it s worthy to note here that we did use one of the temporary stowage bags as a trash container before we got into lunar orbit, and then dumped that stuff in the LM. On the way back, we had two other trash containers. I think most of this was because of our lack of confidence in how much we could put in the waste stowage container.

18.6.4 Water Glycol System

18.6.5 Suit Circuit

18.6.6 Gauging System

18.6.7 Waste Management System

COLLINS   I think there s got to be a better way of fecal containment or disposal. I d like to talk to the experts on that some time later, apart from this debriefing.

18.7 TELE-
COMMUNICATIONS

COLLINS   I thought communications in general worked very well. I think the problems that we had were ground switching problems. We did have extended periods of time without communications with the ground when we were in line of sight with the ground. I don t quite understand all the reasons for that. I m sure somebody else is worrying with the problem more than we are. The VERB 64, I thought, worked well. The only trouble with 64 is that it is a continuous computation. It ties up the computer to the extent that they ve designed it where you have to be in P00 to read VERB 64. I m not sure that if you had to do it over again that would be the best way to design it. I don t know why you couldn t have VERB 64 available in P20, for example. But, in general, all that worked well. Again, we should note that we left the circuit breaker out for the television on panel 225, and that made the onboard tape not available during entry.

18.8
MECHANICAL

COLLINS   Tunnel, probe, drogue, lighting, all worked beautifully. We inadvertently activated the lower left-hand strut softener prior to launch. It was done when hand-controller number 2 was moved and pulled the fabric line that attaches to the strut. We recommend that the backup CMP, or whoever is in there, understand how to reset those strut softeners.

19.0 LUNAR MODULE SYSTEMS OPERATIONS

19.1 GUIDANCE
AND
NAVIGATION

ALDRIN   The dimmer control was adequate. You could tell, as was commented on previous flights, that in the dim range, there isn t a wide range of control. I felt that (for the service alignments anyway) what was available was adequate. I was a little disappointed in the ability to focus the reticle into a sharp image. I thought we d be able to get that a little sharper. It didn t seem to be quite as pressed a reticle as I was able to get in the simulator. I can t explain why that was. The rendezvous radar worked as expected, or better. I thought the signal strength of Borman s on the side lobes was just what we expected, very close to what the simulator depicts, in rise and drop-off.

ARMSTRONG   We ve commented already on the inaccuracy of the simulator line-of-sight needles. That should be changed. I only have one discrepancy on the landing radar - that of the alarm on landing radar position - that we couldn t explain.

ALDRIN   That s probably a computer problem more than a radar problem.

ARMSTRONG   Yes.

ALDRIN   I think the zero Doppler effects down around a 100 feet when the noise came on and then went back off again should be fairly well documented. It was a rather brief period, on the order of maybe 5, 6 seconds, that both altitude velocity lights came on. Then they went back out again.

ARMSTRONG   In fact, I think the zero Doppler drop-outs were less than we expected. Concerning the computer, anything on this 56 or 57 here?

ALDRIN   The overloading of the computer is pretty well understood. It s unfortunate that, because of that, you are not able to take advantage of the use of the radar to designate during ascent. The DSKY, the keyboard that is, we managed to wear out in the simulator. Quite frequently, it would require depressing the keys several times before the entry would be accepted. But the flight keyboard worked very well. I didn t notice that it required any unusual amount of force to get any of the key strokes to take. AGS seemed to work extremely well. The one lighting failure that we previously mentioned in the middle character; the upper left-hand vertical stroke was not lit. This was noticed on initial checkout when a 3 or 9 would come up. Because this particular one was blank, it wouldn t look like any particular number. By filling in one or the other, you could make it either a 3 or a 9. Anyway, there was some possibility for confusion. However, it didn t appear critical.

19.2 PROPULSION SYSTEM

ALDRIN   It seems to me that, in monitoring the gyro calibrations addresses, I did notice that one of them increased to a larger number during the calibration than it finally settled out at. I ll just ask that question of the Systems people when they get there.

ARMSTRONG   We had no abnormalities with the descent engine or ascent engine. We ve commented already on the ascent source pressure; that is, confusion with respect to whether or not both tanks had pressurized.

19.3 REACTION CONTROL SYSTEM

ARMSTRONG   No problems.

19.4 ELECTRICAL POWER SYSTEM

ALDRIN   Everything worked just as expected. The monitoring, the displays were quite close to what we ve seen in the simulator. One small point - in checking the ED batteries in the simulator, when you d push the spring loaded switch, there is some delay before the meter gives you the battery reading. In the spacecraft, this is not the case. It s almost instantaneous. When the switch is placed in either A or B position, the reading comes up immediately.

ARMSTRONG   And there were no difficulties with the ED systems. All worked well, within our ability to monitor its operation.

ALDRIN   In the explosive system, we could hear most of the explosive devices when they were actually fired. I can t recall any that were used that were not audible in the cockpit.

19.5 ENVIRONMENTAL CONTROL SYSTEM

ARMSTRONG   We ve-discussed the CO_2 sensor abnormality and we ve discussed the water in the left-hand suit.

ALDRIN   We also mentioned the slight delay we had in getting the replacement primary canister in. It was a question of not being able to rotate it properly. Then the cap would not go on and completely lock until I was able to just jiggle the canister and get it to insert and rotate properly. It was typical of the sort of difficulties that many people have been having with the cartridge replacement in both primary and secondary. We talked about the temperature levels in the suit on the surface. We explained that rather thoroughly. The temperature got cold before we realized it and, by the time we did, there wasn t much we could do about it to warm it back up in the cabin.

ARMSTRONG   Water supply problems. Concerning the suit circuit, we ve talked about the water problem. I believe, in retrospect, probably the secondary water separator did, in fact, successfully keep the water out of the left-hand suit after about 15 minutes of operation.

19.6 TELE-COMMUNICATIONS

ALDRIN   No comment other than its being a little unwieldy in switching from HIGH GAIN to OMNI just before LOS and then picking up communications again coming on the other side. It seemed to work quite well, with the exception of the attitude we were in approaching powered descent. This produced many drop-outs and required the use of manual reacquisition. Actually, the AUTO would not stay on for a good bit of the face-down portion of the powered descent. I had to make adjustments manually in both pitch and yaw to keep the signals turned on. On the recorder, I think that the previous flights did not spend enough time in the LM to be concerned about the capacity of the recorder, whereas we were going to be in the LM for approaching 30 hours; this being three times the capacity of the recorder. So we did have to attempt to devise some system of turning it on and off depending on our needs. It seems to me that what is really needed is a separate recording system in the LM (and, for that matter, in the command module) that is voice-operated, that turns itself on upon receiving the first signal and turns itself back off again. There s no need for any crew activity to turn switches on or off. It doesn t depend on the operation of the individual audio centers.

20.0 MISCELLANEOUS SYSTEMS, FLIGHT EQUIPMENT, AND GFE

20.2 CLOCKS

ARMSTRONG   Want to comment on the LEB mission timer first?

COLLINS   In the command module, the LEB mission timer ran slow. The first time we checked it after lift-off, it was 10 seconds slow. We reset it to the NOUN 65 value, and the next day it was 3 or 4 seconds slow.

ALDRIN   I think it slipped a digit. This was probably the cause for its being off.

COLLINS   No, I think it just ran slow. The digit that slipped was in the tens of hours digit. For example, when it was supposed to be 134 hours, it was reading 144 hours. That happened, though, after you guys left the lunar surface; and I m sure it is as a result of my having manually set it to LM nominal lift-off time.

ARMSTRONG   I don t know how that thing runs slow.

COLLINS   It did indicate that

ARMSTRONG   It never agreed with the other time; it was always off by a varied amount, on one digit or another.

COLLINS   In addition, there was a small crack in the glass

ARMSTRONG   Of course, the LM mission timer failure on touchdown was documented during flight. The next day, we were able to reset the time with it; a 30-second reset.

ALDRIN   I think it s worth noting that I feel that it s extremely unfortunate that we don t have clocks that count down to zero and then reverse and count back up again. It s forced on the crew anyway. I think other crews are doing the same thing; setting clocks to count up to a burn and then reach 59 and 60 at ignition, just so that you will have a clock that s counting up during a burn and post-ignition time. I don t expect this to be done immediately. I d sure like to have it recorded for posterity that clocks would operate much better if they counted down to zero and then back up again without requiring the throwing of a switch at a critical ignition time to get the clock to do that.

20.6 CLOTHING

COLLINS   The constant-wear garment looks to me like something that has more work put into it than it really deserves. The results are less satisfactory than they d be if we just went flying in our regular old summer flying suit. Summer flying suits have more pockets in them and more places to stow things. They are garments in which we feel completely at home and they are more comfortable than the two-piece CWG. If the CWG has to be made out of a fireproof material, then regular old summer flying suits can also be made out of that same material. It would save considerable money just to delete the custom-tailored CWG and let us pack summer flying suits made of the appropriate material.

ARMSTRONG   I think we ve discussed the BIG in some detail during recovery.

ALDRIN   Under coveralls, I think both you and I noticed a slight itching in the forearm. It was probably just a question of wearing through the Teflon liner.

COLLINS   I d like to go back just a second here and interrupt. I wasn t talking about the constant-wear garment. I was talking about, I guess, the flight coveralls. I can t even keep the names of them straight. The two-piece white jobber that you wear. The underwear I was not criticizing.

ARMSTRONG   I agree with Buzz. There was, near the end of the flight particularly, some irritation of the forearms and elbows, which I think is reaction to the Fiberglas. We noted this before in altitude-chamber runs and so on; and concerning the Teflon-coated garments, I think that it was just a breakdown through continued wear that exposed the Fiberglas through the Teflon and caused a typical reaction.

20.7 BIOMED HARNESS

ARMSTRONG   Now in the BIOMED harness area, we had a few discrepancies.

ALDRIN   Yes, I had two that I d like to note. Both of them have been documented. The center chest lead dried out, and I was requested to make a change in that, which I did. The right lead on the right side of my rib cage evidently rubbed against the suit and caused a minor laceration on the aft part of my side. I don t know if there is any way around that other than just not wearing those things.

ARMSTRONG   With respect to sensors and harnesses causing discomfort, from about the middle of the flight on, the sensors were essentially itching. I had a tremendous desire to scratch them off.

COLLINS   That s right.

ARMSTRONG   I scratched all around every sensor about a thousand times. That s just an inconvenience and a distraction.

COLLINS   I think part of it has to do with shaving your chest and then the hair starts growing back underneath the plastered down sensor. That was the impression that I had of it. The ones that itched the most were the ones with the most hair around them. They left little marks that went away in a day or two.

ALDRIN   Let s digress to the lightweight headsets. I ve found it preferable to use the lightweight headsets instead of the COMM carrier. It still didn t fit too well on your head. The mike boom and its attachment to the headpiece just doesn t seem to be the best arrangement that could be worked out. It s a lot superior to others that we ve tried, though.

COLLINS   I take the lightweight headset apart. The piece that goes around over your head I throw away. Then I attach the microphone to my collar somehow with an alligator clip. I take the long-eared tube and tape it to my ear with a piece of adhesive tape. That s the only way I can stand it. If I put it around my head, it drives me crazy after a couple of hours; not to mention falling off all the time.

ARMSTRONG   The only difficulty I noted, because I like the lightweight headset myself, is the fact that the mike boom and the head band are a quick-disconnect arrangement, which is continually disconnecting. I m sure there was some good reason for that. It may be desirable for some people, but I would much rather just have that thing firmly connected in the proper location and leave it like that.

ALDRIN   I used the molded ear-piece inside the COMM carrier in the LM through activation and power descent, but it became so uncomfortable that, after we were on the surface, I removed them and continued the remainder of LM operation without them. They did increase the volume during the time that I had them in, however.

20.10 RESTRAINTS

ALDRIN   I think I ve already mentioned that, in the LM, the LMP s restraint system tended to force you forward and to the right and required leaning back to the left to maintain balance. This was a little bit disconcerting.

20.15 CAMERA EQUIPMENT

ALDRIN   The only possible malfunction that was observed was with the LM 16-mm camera. Evidently, it worked properly, but it didn t seem to give the proper indication. Initially on connecting power, the green light came on and, after 10 to 12 seconds, it went out. However, once we started taking pictures subsequent to the initial turning on of the camera, whenever the power came on, the light came on and stayed on throughout the time the power was applied to the camera. So it didn t really give an indication as to whether frames were being taken except when you observed the light to be blinking.

21.0 VISUAL SIGHTINGS

ARMSTRONG   Most of the items in Section 21, Visual Sightings, have been previously reported.

21.4 TRANSLUNAR AND TRANSEARTH FLIGHT

ALDRIN   There was only one minor observation returning from the Moon. Looking back at it, at a time after Mars had passed behind the Moon, there was one time period where I imagined that the image of Mars was coming from a region where it couldn t come from, because it was in a dark portion of the Moon. This obviously was an optical illusion of some sort.

ARMSTRONG   I suspect that it was, in fact, just immediately adjacent to the horizon.

ALDRIN   We must have looked at it immediately after it had come from the back side.

ARMSTRONG   Yes.

21.5 LUNAR ORBIT

ALDRIN   In lunar orbit, following ascent, we did note and mention to the ground that approaching CDH when the Earth came up above the lunar horizon, I observed what appeared to be a fairly bright light source which we tentatively ascribed to a possible laser. That seemed to be the best possible explanation until we were coming back in the command module approaching the Earth and were able to observe something that gave about the same appearance. When putting the monocular on the light source, it appeared as though it was the reflection of the Sun from a relatively smooth body of water such as a lake. I think we ve revised our initial conclusion as to what the source of that light was that we saw coming from the Earth. If no one owns up to having beamed the laser toward the Moon at that time, it was more probably a reflection off a lake. I still think it s an unusual phenomenon, at that distance, to see so bright a source of light. In the film, it didn t appear as though this was going to show up at all. The Earth was too bright.

22.0 PREMISSION PLANNING

ARMSTRONG   First, we can say that essentially the entire flight was flown on the mission plan and the details of the flight were, in fact, in accord with the flight plan.

ARMSTRONG   The flight plan was really very well written, and we found very few discrepancies in flight. In terms of system operation, normal housekeeping chores, pre and post, sleep, checklists, burn reports - all those things were included; we very rigorously followed all the instructions day by day in all 135 pages.

COLLINS   A good flight plan; a lot of hard work went into it.

ARMSTRONG   Spacecraft changes were relatively few in the final stages, although there were numerous replacement items. Fortunately, there were not too many configuration changes in the preflight phase.

ARMSTRONG   There were relatively large numbers of procedural changes filtering through this system daily, right up until and during flight. Some of these procedural changes were relatively significant; others were very small. I can t say how it compares with other recent flights in the same time period, but it was our impression that the procedural changes were excessive and indicated that generally we hadn t completed our preflight planning as well as we would like to have done and in timely enough fashion for a mission of this consequence.

ALDRIN   What this means is that a good bit of the training had to be developed. It had to be devoted to the development of these procedures in the new areas that we had in our mission.

COLLINS   I think, in general, if the crew wants to make changes to their procedures, they should be discouraged from making any unnecessary changes. The time period when the crew really should be in the change loop, I think, is fairly early in the training cycle, a couple of months before launch. During this time period, from the command module view point, I found it difficult to promote changes. I found that there was a considerable time lag between my requesting a change and my seeing a new checklist or a new rendezvous procedural page or what have you. It ran on the order of several weeks. This is several months before the flight that I m referring to. This is the time when I was trying to get the rendezvous procedures optimized. In the checklist world, I was trying to get obvious mistakes corrected, trying to do this early in the game. It was very discouraging, because the results frequently came back not exactly with the changes as I had intended them; and there was a considerable time delay between the time that I requested the change and the time that I saw a new piece of paper in my hand - on the order of perhaps 3 weeks. Now late in preflight, everybody got all hyper and they got streamlined. A week or two before the flight, when I desperately wanted not to make changes, the system was all for it. If changes were required, from the day you requested it to the day you had a new piece of paper in your hand was more like a day or two. Now, that s the kind of service that we needed months before. It would have saved a lot of man-hours of work in the long run to have a fast-response system early in the game; let all these changes reverberate throughout the system, go to the contractors, and come back. It ended up that when the quick response was needed it was not there and late in the game, when we didn t want to make changes, everybody was hovering around us saying, Well, how about this? Do you want to change that? Do you want to change the others? In general, it s the same thing I think we had in the Gemini program; that is, in the final phases of training, we really had superb support and help. If you could drain off a little bit of that and give it to that crew earlier in their training cycle, I think things would, in the long run, be a lot more efficiently handled. It s a case of not having enough help early and having too much late in the game. Well, I m sure I only see a very narrow little section of the total operation. Just from my parochial view point, it would appear that if you have a glob of help this big, instead of putting all of it on the next flight to fly, you take a little chunk of off and give it to the following flight.

ARMSTRONG   You wouldn t need so much help at the end. You wouldn t have such frantic finishes.

ALDRIN   I think we ve already commented on a specific change in the mission profile; namely, the change in the CSM orbit from circular to elliptical and the two effects that had that I don t believe received enough preflight attention. One of them was the change in the radial component of the CDH maneuver from a small value to a significant value. Our burn had an 18-ft/sec component. The other effect was the range-rate values that we had despite the fact that it was a nominal insertion and a nominal trajectory approaching CSI. The range-rate values were outside the limits of the backup chart, rendering that solution useless.

22.5 MISSION RULES

ARMSTRONG   I think that we had a good working relationship during the formulation of mission rules; essentially those rules are worked out at the working level sufficiently early that we had very few head-knocking sessions on disagreements on the rules. We flew with rules that were generally agreeable to everybody.

ALDRIN   I think the Flight Operations people were most cooperative in working these things out and explaining the various peculiarities to us.

ARMSTRONG   It is somewhat of a problem for the crew to know the mission rules well enough to fly the flight according to rules. It s a very extensive and detailed document, and, fortunately, we didn t have any trouble. In time-critical phases, it s quite a problem to recall all the rule combinations that, in fact, have been carefully decided preflight. We used some gouges and did take some short, streamlined versions of the mission rules (the significant ones) on both the CM and the LM, should there have been any problem when we did not have COMM available to discuss the situation.

23.0 MISSION CONTROL

ARMSTRONG   There s very little comment there. It worked well. Data transfer from the spacecraft to the ground worked well. We were well advised of our status and consumables and so on. Updates went as per simulation. There were some real-time flight plan changes during the flight which I think always could be accommodated.

COLLINS   I just thought, in general, that we got outstanding support from all four of the teams involved. I thought that, in general, everything was beautifully worked out, and I don t think we ever really had any serious disagreements before the flight or during the flight.

ALDRIN   Well, I think the CAP COMM s are to be highly commended for their very detailed understanding of every particular phase of the mission and just what was going on inside the spacecraft. I think they did an outstanding job.

COLLINS   I do, too. I think they did a superb job and so did all the rest of the team. I thought we had just outstanding support. I couldn t be happier.

24.0 TRAINING

24.1 CMS

COLLINS   I think the CMS, in general, was an excellent simulator. Its weak point is its visual system. Some improvements were made during the course of our training. For example, some of the Apollo 10 photographs were put into the sextant and telescope visual to enhance P22 training. Various adjustments were made to the transposition and docking, the window display, the model was tweaked up, and so forth. I d have to say, though, still in general, that the visual is the weak spot in the training, and P22, P23, transposition docking training suffers because of it. The crew station was well equipped. It was brought up to the 107 configuration after the flight of 106. We inherited all the various stowage compartments, or most of them I should say, so that during the last 6 weeks or 2 months of training, the crew compartment quite closely resembled the interior of the spacecraft, closely enough. If I had any changes to make to the CMS, I would spend the money on trying to improve the visual simulation. I think the people

who work most intimately with the CMS are those that are most aware of its visual shortcomings. They understand the changes that have to be made. It s just a question of getting the money pumped into the system to make the necessary design changes. The availability of the CMS was quite good. It did bomb out a time or two, in which event the CMS 3 was usually made available for our training. I think that the crew of the spacecraft next to fly should have the right to use the other simulator to keep the mission simulation schedule on an even keel. I don t think you should just arbitrarily kick the next crew off on any old day when your prime simulator bombs out. However, during times when Mission Control  is on the line, I think either simulator should be made available depending on which is working better for the crew on the next vehicle to fly. That s about all I ve got to say about the CMS.

ARMSTRONG   Well, I d like to make a couple of comments here, and it applies both to the CMS, LMS, and all our simulations in general. First, I d certainly agree that the strongest shortcomings of both the simulators is the visual. It s able to do a reasonable job on the stars, star patterns, and the things necessary for optics for platform alignments. Beyond that, the abilities of those half-million dollar window extravaganzas is negligible.

ARMSTRONG   There are a lot of areas that could very well stand an improved visual simulation for training.

ALDRIN   I think one of the biggest drawbacks of the visuals is the lack of illumination coming into the spacecraft from the window from either the illuminated surface, or more important, from the effects of the Sun shafting coming in. We re always operating in much darker conditions inside the simulator than those that actually existed in either the CM or LM.

ARMSTRONG   Now, the second area that suffers in fidelity is probably less important, but it s factual. That s the area of all the gas and fluid systems in the spacecraft. Our simulators do a good job of electrical electronic simulations but do an absolute zero job when it comes to what do valves do; not in how they reflect in the gauges, but how they actually affect gas flow around the spacecraft through valve sounds, water flows, and how to operate those devices. Apparently it isn t mandatory, because we managed to fly the spacecraft and operate those systems without ever looking at a simulation of them. I think it is a fact that we do not have a simulation of anything like the glycol loops, or the suit loops, the effects on suits, of operating valves in the spacecraft in both CMS and LMS; things like REPRESS, cabin circuits, and so on are just absolutely not represented except as they are reflected on gauges, which really is a relatively small part of the operating of fluid and gas systems. You want to talk a little bit about the switch?

ALDRIN   There were several changes made between LM-4 and LM-5, and I think the LMS was a little bit late in getting some of these modifications such as some changes in circuit breaker locations, the LCG pump circuit breaker, the radar GYRO switch. We did get these in, I m not sure the exact timing something on the order of 3 weeks, something like that, before flight. It would have been nice to have had that package of modifications completed sooner.

ARMSTRONG   They were installed essentially coincident with the time period when you should essentially have simulations completed and just be brushing up on that one. I think, generally, the availability of both simulators, with the exception of a few days when they were unavailable, was good. We could generally depend on having one of the two CMS s and the LMS available to us for training. Very seldom did we have to sit around and wait for the simulator to be ready. The fidelity wasn t as good as you d like, but their availability was good.

COLLINS   A high degree of availability was absolutely mandatory on our training cycle. If we had poor to bad luck with the simulator availability, I don t believe we could have flown the mission of the 16th of July. I think it would have been unwise for us to attempt flying the mission on the 16th of July with much less simulator time than we actually had.

ALDRIN   What we re really saying is that both the LMS and the CMS were the key items of training, and so much of all that we did depended on their operation.

COLLINS   That s right. When the system really committed to a July launch, I m not sure when that was, but my impression was that that decision was made fairly early, although not officially.

ARMSTRONG   It was soon after the 10th.

ALDRIN   Yes.

COLLINS   When that decision was made, a very vital part of it was the fact that the assumption was made that those simulators would work properly and that we would have a high degree of availability for the remainder of our training cycle. If we had stubbed our toes a few times along the way, I don t think we d have been prepared to launch in July. I think we were lucky. The simulator availability was up, and we were able to grind away hour after hour of good fruitful training, particularly in the mission SIM s with MCC.

ALDRIN   I think the visual - we may be knocking it a bit too much and not pointing out some of the good features. I think it did quite a good job in the pre-PDI observation out the window of making uses of films that were taken on Apollo 10. I thought this was incorporated into the system in a very, very efficient and very well reproduced fashion.

ARMSTRONG   The new L&A, with the new model of landing site 3, certainly was a gigantic improvement of the previous lunar surface visual.

ALDRIN   Yes. Unfortunately, it lost its usefulness at about 100-foot altitude. Isn t that right?

ARMSTRONG   Yes.

ALDRIN   So it could not be used for the very final manual phases of the touchdown. Again, most of the final phase of descent was restricted to the presentation made in the left window, the one that was in the right window was not a correct presentation, because it essentially was the same view that was out of the left window, which put it off the angle by some 50 or 60 degrees. Now, it would have been nice to have had a better visual presentation in the right window. I wouldn t put it in the mandatory category because at this stage of the game I don t think that the roles employed by the crew members required a high-fidelity visual in the descent phase for the LMP. I think his tasks were more occupied monitoring onboard systems, relaying the information that was displayed by the computers, and by the radar system as an assistant to the Commander.

24.3
INTEGRATED
SIMULATION

ALDRIN   When they worked, there was no doubt that they were extremely valuable. We did lose a fair amount of time because of computer problems; not the spacecraft computers, but the computer that ties the two together.

ARMSTRONG   In this type of a mission plan, of course, the integrated simulations are a very vital part of the training and not just the training. They are a vital part of the procedural development and checklist development that s required to gain the confidence level that you have to have to begin a flight of this type. In general, I think that we performed a lot of integrated simulations, and they were, in general, very beneficial.

COLLINS   I don t think we performed too many, I think we had about a minimum number. I think, perhaps, we spent a little more time than we should have on launch/launch-abort SIM s and maybe not quite as much as we should have on lunar ascent SIM s.

ALDRIN   Yes. I agree that it was satisfying minimum requirements, and we certainly could have used up to double the number of integrated simulations.

ARMSTRONG   In general, the simulations where both vehicles were airborne were pretty good. Simulations where we had one on the ground and one on the surface were probably less productive and less like the real case. I doubt that P22 s and things like that ever really worked well enough in the simulator to give you a good understanding of that part of the problem.

COLLINS   Yes. That s true.

24.4 NETWORK SIMULATIONS

ARMSTRONG   When we talked about integrated simulations, we ve been also talking about network simulations because far and away the majority of our integrated simulations were performed as a part of the network simulations. As a matter of fact, in the final months, prior to launch or perhaps six weeks prior to launch, the great majority of our time was spent on network simulations. Such a large percentage was spent on simulation with the network, as a matter of fact, that we had difficulty finding time to do simulations that were not covered by the network simulation.

COLLINS   That s true.

ARMSTRONG   I would guess that about 60 percent of the days were covered with simulations with the Mission Control Center.

COLLINS   Again, this is the sort of thing where we seem to fight our way through these flights one by one and the SIM s come very late in the training cycle. It would appear to be very valuable that some of these first SIM s could be moved up in time so that we don t have this last minute cluster of them. It makes you nervous from a number of viewpoints. First, you don t really have as much time available to putter around with those things which you know you re rusty on and you need training on. Second, the monkey s really on the back of all the electronics people to keep those things running, and you really have to keep them running, and you ve got to go through successful day after successful day or really you will not meet what you set as minimum training requirements. I think there is a lot of pressure there on a lot of people at the end to keep those simulators running. There s pressure on the crew and on the maintenance people, and the pressure all around because we delay those SIM s too late in the game. The reason we do delay them is that the total system, the Center, is really only capable of facing up to one flight at a time.

24.5 DCPS

ARMSTRONG   The DCPS was used only for launch, launch abort, and TLI simulations. Although we were unable to afford the time to do many of these as we liked, the simulator worked reliably, and the procedures developed there were very useful. I think it s probably appropriate to say that as the flights have progressed, the launch abort procedures have fortunately become more and more streamlined and simplified. I think at the present time they are in very, very understandable and rational form. It s an easy job for the left-seat man during launch to understand those procedures and be able to implement those irrespective of other situations.

24.6 LMPS

ALDRIN   Well, I was a little disappointed in the variety of abort cases that the LMPS could handle. We looked at the DOI aborts. Primarily, it was a rendezvous trainer. Starting at insertion and for nominal rendezvous cases and then a good bit later in the game, we were able to pick up certain selected abort cases. But, as I say, I was a little disappointed in not being able to run through a wide variety of different cases. There were potential cases that could have arisen.

COLLINS The good thing about the CMPS was that it could investigate a number of dispersed rendezvous cases. The bad thing about it was, first, as far as Apollo 11 was concerned, we got caught in the middle of a move from one building to another, and the timing was extremely poor. It made the CMPS not available early in the training cycle when we could have made very good use of it. At that time, the CMS s were really not available. Later in the training cycle when the CMPS became available, it was of little or no value to the crew itself. It was of value to other people, but to the crew itself was of little or no value, because we were then spending all our time on the CMS and had no time to devote to the CMPS. So it was just a masterpiece of poor timing for Apollo 11, and Apollo 11 comments I am sure won t carry over to other flights. The bad thing from a technical viewpoint about the CMPS is that the computer is a sort of an idealized simplified mechanization of the real computer, and it was always several iterations behind the latest MIT math flow. It never quite worked like the real computer worked. I guess during the rendezvous this is an important factor.

ALDRIN I think the same thing applies to the simulation of the LGC. It was considerably limited, and there were many little tricks that we employed that you could exercise in the LMS. When you try and work them in the LMPS, the computer wouldn t respond properly, and to that degree, there was a certain amount of negative training because we d have to establish other procedures, other work around techniques to come up with the same information.

COLLINS I d like to emphasize that what I ve said about the CMPS and its shortcomings is not the entire story. I was talking about the actual crew training on the CMPS. Now, above and beyond the crew training, the CMPS was used as a procedures development simulator, and the McDonnell Douglas people spent a lot of hours looking at various trajectories and various dispersed cases and also on various abort modes. The thing was of great value in putting together my solo book with all the dispersed cases and all the abort possibilities. So I m sure it was of great value. It was just that because of poor timing it was of very limited value as direct crew training.

ARMSTRONG LMP and CDR had participated in centrifuge training in preparation for Apollo 8 and chose not to repeat that.

COLLINS I thought that half a day on the centrifuge was useful. I don t think it s mandatory. I think you could go ahead and fly the flight safely without any centrifuge training at all, but I think that to do a couple of entries in it is worth the time. It s well worth the time spent there.

ALDRIN   That was good that we had done that on Apollo 8. I thought that was well worth the time.

ARMSTRONG   The launches were never simulated in either case and I don t see any requirement for launch acceleration.

COLLINS   The only hooker there is if you get into these abort cases where you re pulling horribly large numbers of g s for great lengths of time. We just ignore those, and I suppose that s probably all right. Somewhere in all our background, we ve had Johnsville centrifuge runs up to 15 g s and things like sustained peaks over 10 g s, and I m not sure that the crews that are coming in now have had any experience like that. I think one time is worth it to see a very high g spike and to see a fairly long period of time at a moderate g of 10 or thereabouts. There s no doubt about it that there are certain little tricks about breathing that would be nice to know and to remember in the unlikely event that you did have one of these high g aborts. So I d say that a general background training is worthwhile. Having that under your belt, I think half a day for a specific mission training of entry would be more than adequate.

24.9 THE DOCKING SIMULATOR

ARMSTRONG   The docking simulator was used very little. Some simulated LM dockings were performed in that simulator, and in as much as that is a secondary docking method, we felt that was adequate. Had I more time available to prepare, I probably would have spent somewhat more time on that than I did. As mentioned earlier in the debriefing, the shortcoming of that simulator is that it doesn t provide any of the simulation of post-contact dynamics. That is, of course, the area where we ran into a little problem, so scheduling of that by the training people is probably warranted.

COLLINS   I think the docking trainer should be command module active rather than LM active. I don t know how much it would cost to convert it, but the thing is going to sit over there and cost money for its upkeep and people to run it and all that. It s probably worth a little extra to make it command module active rather than LM active.

24.10 RDS

COLLINS   Well, I feel the same way after the flight that I did before about Langley. That is, if Langley is up and running, it s well worth the trip to Langley to make use of it. But I don t really think that you can put that simulator into mandatory category. Once it s dismantled, I would say leave it dismantled. There is no firm requirement; it s not mandatory to look at the Langley simulator. It s useful, it s real, it s full-size, it gives you a good out-the-window display. Although it is hydraulically operated, its control system response is very close to the real thing, and if it were in existence, I d sure take advantage of it again as I did in the past. You guys flew that, didn t you? What do you think?

ALDRIN   No, I didn t fly it.

ARMSTRONG   Not recently.

24.11 FMES

ARMSTRONG   We didn t have the opportunity to participate. I m quite sure that there were many areas of interest, that would have been valuable for the crew to look at in the FMES. We could not afford the time, and with the unpredictable schedule of the

FMES, it was just impractical for us to try to incorporate that into the training period.

COLLINS   That s the same with the North American evaluator. I tried up until several weeks before the flight to find time to go out to the evaluator and look at sort of a summary of what they found in their Apollo 11 verification work. But it just didn t work out, and I put it in the nice to have category rather than the mandatory category. I think you d learn some things from it, and it would be a good crosscheck on the CMS, but I don t think it s mandatory that the crew participate in the evaluator.

24.12 EGRESS TRAINING

ALDRIN   I don t feel that the tank egress exercise is worth the time. It seems to me it can go right to the Gulf. We understand the procedures well enough, and there are no difficulties from a safety standpoint that really warrant exposing the crew to both the tank and the Gulf.

ARMSTRONG   The Gulf exercise is relatively productive for the amount of time it takes. It s a half day exercise, and I think that it was well organized for us. For the couple of hours work that it takes, you probably get a good bit of confidence in your ability to handle the spacecraft in the water, and you re obliged to do something in that regard.

COLLINS   Yes, I agree. You probably delete the tank, and I think, oh, it s a pleasant ride out on that boat. You re sleeping or sitting around doing nothing. I think probably if you re going to get the most out of it, you could precede the Gulf work with a couple of lectures and briefings on the boat on the way out on what you are going to do and how to run through all this stuff and the whole schematic on the post-landing events system and maybe a few words on the survival kit. If you delete the Stable II training in the tank, then that time could be spent on briefing, on some other precautions, such as the no-no s involved in Stable II egress. I think that would be a more productive day if it were arranged that way. On the other hand, I enjoy sitting around in the boat. It s relaxing.

ARMSTRONG   Pad egress and mockup egress have to do with preparing for an emergency in the spacecraft on the pad. This is required for the chamber operation for safety in the chamber. By the time you get through that, not very much additional is required for the spacecraft, itself, on the pad. I don t think we spent excessive time in that area, though.

ALDRIN   I do think you want to accomplish some of these exercises with flight gear because the training suits are used so much, and fasteners and connectors always seem to work a lot easier. It s a good exercise to run through at least once under requirements to move quickly with the flight hardware.

24.13 FIRE TRAINING

ARMSTRONG   It s valuable, if you do have a fire and have to use it. Other than that, it s like buying insurance. If you never use it, of course, it s just a time expenditure that was nonproductive. I ve no objection to that fire training.

24.14 PLANETARIUM

ARMSTRONG   We had a very limited planetarium training exposure. It s primarily due to the fact that we just didn t have the time to look into it in more detail. We had relatively extensive planetarium flight related work on previous flights. Our work was limited strictly to the flight plan itself and work in the flight: things that could be learned in the planetarium that would apply to the mission plan, stars selected for alignments, what could be seen from the lunar surface, what constraints due to certain lighting conditions and locations of planets, those sorts of things. We did it relatively early in the training cycle to get a basic understanding of what our geometric situation was, and that was probably worthwhile.

ALDRIN   I think it was. It s useful in a general sense, because you might be able to get some early information on the planets. The simulator doesn t have any representation of the planets. I think most of the specific navigational use of the stars and the star field can best be done with the existing CMS.

COLLINS   I think one trip to Chapel Hill for a training cycle is useful.

ARMSTRONG   That s what I think. The geometry for fixed mission launch time is pretty well fixed on a lunar mission. That is, once established in a launch date, the entire astronomical geometry is fixed. A good understanding is very useful. It influences a lot of things later in the procedural developments, so that session is probably worthwhile. It can be improved by having a mission planner that understands the geometry and the constraints of this launch date involved in the planning of that training session. We tried to do that, and we really didn t get as much out of that aspect of it as we would have liked.

24.15 MIT

ARMSTRONG   I want to talk about MIT and, in general, in our flight, this was restricted to understanding of the programs and program changes, as in the software end. The hardware was pretty firm at the time and we all had a fairly good understanding of the hardware prior to this training cycle. The understanding of the program and program changes, however, is one that unfortunately takes a lot of time. It takes a number of separate sessions in smaller groups throughout the cycle. I really think that rather than one big 3-day session of a review of the programs and so forth, it would be better to have a number of smaller sessions interspersed at various times in the training cycle so you could limit yourself to just one phase of the mission at a time.

COLLINS   I spent as long as 5 days, not on this flight but on others, sitting in a chair at MIT listening to a chronological description of the math flow, and it drives me crazy. By about the second or third day, I m just saturated; it s like filling up a teacup with a fire hose. There has to be a better way of doing that than just going to MIT and sitting for days. Maybe if they are broken up like you say, it will help.

ARMSTRONG   This is a difficult area because various people assimilate information differently. I get absolutely nothing from 2 hours of going through logic diagrams, while other people find that very informative. I much prefer going through the operator s checklist and trying to understand each step in the operating checklist, what that does, what information is being displayed, and how it s being processed, than going through endless software loops on a diagram.

ALDRIN   Yes. Until you re related to specific use situations, it doesn t mean very much.

ARMSTRONG   You have to understand the basic thing and go do it in the simulation; then you can understand better some of the details, why it s done this way, and what options are available to you. Just to spend long days in reviewing hoards of GSOPS and things like that, in my view, is a very unproductive session.

ALDRIN   We ve gone over programs such as P20 many times, and I still can t recall all the logic flows of different paths that the computer is taking in acquiring radar lock-on. I think, in a sense, our checklist may suffer somewhat in that it doesn t help too much either in this respect, in covering the various ways that you handle abnormalities in acquiring radar lock. Certainly, the time spent in going over all the logic flows wasn t particularly productive.

24.16 SYSTEMS BRIEFINGS

ARMSTRONG   In general, we didn t have long courses of systems briefings. We chose to have an expert on a particular system come in periodically and review that system on an available basis. This worked all right, but in my own case, I felt that by the time we got within 2 months of the flight, we still didn t understand some of the systems or hadn t gotten around to understanding them in the depth that was required. I don t know how to get around this. This was the problem in our flight that was just due to the very tight training schedule that we were on. There just wasn t time to do all the things in the order and in the depth that you wanted.

COLLINS   I have one concrete suggestion for these systems that centers around CMS-1. CMS-1 is the simulator where the crews get most of their systems training, because they re on CMS-1 before they go down to the Cape and get the more mission-oriented training. CMS-1 has some very good instructors. It uses a different system than CMS-2 and 3. CMS-2 and 3 have people who are trained across the board in a rather shallow fashion. CMS-1 has people who are trained in a narrow area but in depth, and they have some good people. One flaw is that the CMS-1 instructors know how the system is designed and of what it is capable, but they have no more idea than a rabbit of how the equipment is actually used in flight. On a few occasions, I have had people from FOD EECOM s to come over and sit in on CMS-1 briefing sessions, and it ended up being more of a briefing session for the CMS-1 instructor than it was for either me or the FOD people. I think there should be some way at MSC to get the right hand and the left hand together, to get the CMS-1 instructors up to speed not only on the basics of their system and its capability but to go one step further and get them into the Control Center, get them to know the EECOM s, and get them to further understand how the equipment is actually used during the course of a flight. Then I think they d be much better prepared to present to the crew those things that are really important and not trivia along with the important details. Maybe there are pitfalls there, but I have the feeling that I got a hell of a lot out of CMS-1. I liked CMS-1 and the instructors. It was sort of the backbone of my systems training. But I think it could have been a hell of a lot better if it could have been integrated into more of a real-world approach. If those people were familiar with the everyday operations of Mission Control; the downlink; what they have on telemetry; how you use the water boiler, not how you could use it but how you really are going to use it, that would have enhanced that training a lot.

ARMSTRONG   We had only a few hours on the launch vehicle, which is probably about right. It s fortunate that, because of the high reliability of the launch vehicle, we haven t had a requirement to know in depth a lot of alternate switches. In most cases, they are available to you anyway.

ALDRIN   Not too much you can do about them.

ARMSTRONG   There are some things, particularly in the S-IVB re-light, that it s important to understand very well. If it works perfectly, it s going to run right in spite of you. But if there are abnormalities, it s very good to understand what the effect of those are. So some amount of time is required there. I think we hit that about right.

ALDRIN   I think the DCPS did a very good job in relaying launch vehicle peculiarities to us.

ARMSTRONG   They did. The DCPS people do a good job in that area. They understand. They have kept very close to the launch vehicle changes, and I think they have really been able to keep us better informed on important things to know in the launch vehicle than our formal Saturn briefings.

24.17 LUNAR SURFACE TRAINING

ALDRIN   I think there were enough uncertainties about the one-sixth g environment to warrant the degree to which we used both WIF and the KC-135. Looking back on it now, I don t think the follow on crews will need as much as we did. I believe that the more productive training would be with the KC-135. It would be nice to have a better simulation of the surface characteristics. That is the big shortcoming, I believe, of the KC-135. But one-sixth g is relatively easy to operate in. It doesn't take too much detailed training, I don't believe.

ARMSTRONG   As far as the use of POGO, I think it's worthwhile. It takes very little time to go over there and train. From the viewpoint of the directorate maintaining the POGO and what its cost of operation is in terms of money, staff, et cetera, I can't say what

the balance is there. If it didn't cost anything, there would be no question that the little bit of time that it takes to go over there and get that kind of experience is worthwhile. It's not mandatory. The KC-135 is unquestionably the best simulation of one-sixth g. It's got certain limitations as we all know. You can't do very much, and it's very expensive in terms of the amount of time it takes you to get a little practice.

ALDRIN   You're probably not going to remember them. I think the best time to do that is just at the beginning of the EVA.

ARMSTRONG   The one-g walkthroughs, of course, were the basis of our timeline planning. I don't see any way of getting away from that. You're going to do a number of those one-g walkthroughs, and you're going to develop your timeline and the procedures. There isn't another way to do it right now that's a good way.

ALDRIN   Yes. That's the only way to do it. You can't sit down.

ARMSTRONG   It's well worth it. We would like to have been able to do a few more had we had the time. I think that we do need to improve our facility for that job. We need to have a better LM, more accurate simulation of the LM. We need a better and larger area to work in. We need more topography and variations of environment to work in so that the simulations can be as good as you can reasonably afford on the ground. I think ours was less accurate in terms of its fidelity than we should have had to properly plan that. We should have as much flight-type equipment as we can in those exercises. It's going to result in an increased productivity of the time you spend in EVA.

ALDRIN   I think the Sun position relative to the spacecraft is extremely important. There are so many things in the EVA that are completely dependent upon the lighting conditions that you have, such as placement of experiments and the photography. We were prepared to do it in one fashion, assuming landing straight ahead. With little variations of that, we generally knew how we were going to approach them, but it was going to be a real-time decision for the most part. With the small amount of yaw that we had, it did perturb our operation to some degree. I think that the one-g walkthroughs ought to look at specific variations in LM orientation and touchdown.

ARMSTRONG   We didn t do much in the way of field trips. We did one geology field trip. We never could afford one which we thought might be particularly valuable for its scientific return.

ALDRIN   I felt for the most part that the trips we went on suffered a good bit from lack of realism. Maybe we just didn t get into this latter one, that we missed, soon enough. I think, based on some of the information that we ve acquired on this flight, we ll be able to make many of the field trips a good bit more productive. I think both SESL and the 8-foot chamber are valuable. It s hard to say that they are required. They do give you that additional confidence in the flight equipment. It would be nice to be able to operate in both of them, for example, with the OPS, and not have restrictions as we had in the SESL in not being able to use it. I m not sure that the thermal aspect of the SESL tests is needed.

ARMSTRONG   Yes. It proved to be non-informative, which I guess was good, because it said we didn t have any problems. I agree with Buzz. The confidence that we got out of that was very good. I m glad we did it. I m glad that we worked with the flight equipment in there and, in a cursory fashion at least, as tests rather than walking up and down on a box. Not that it really taught us all that much. Again, it was just a confidence builder, and I think that we could probably do less in the future. It is important, however, for anybody in a surface activity to have a high degree of confidence in his ability to operate his equipment. That is what that gives you.

ALDRIN   In the SESL, I think it was more important to be exposed to the lighting than to the thermal environment. It was the only place that came fairly close to duplicating the wide variation of lighting conditions.

ARMSTRONG   Our briefings on lunar surface training were more give and take sessions on planning the various procedures and deciding the most efficient way to use our surface time. That s consistent with many other areas of the flight, I guess.

**24.18
CONTINGENCY
EVA TRAINING, KC-
135, WIF, ONE-g
WALKTHROUGHS**

ARMSTRONG   We did contingency transfer in the WIF and went through some procedures with that group. Since we didn t have to use it again, it fits in the insurance category.

ALDRIN   There is such a wide variety of contingency situations that can come up. You can t train for all of them, and I think you have to cut short a few of the available possibilities and just say that if you have to face that one, you re going to take the time and work it out in real time. There are a wide variety of exercises - one PLSS; one OPS; two PLSS, sometimes with OPS and sometimes without; and transfer of hoses through the tunnel. You just can t train for all of them. Somebody has to sit down and try to work out procedures. I think you do need to take a good set of in-flight contingency procedures that will handle the cases that may come up.

COLLINS   You need at least one good long session inside the command module with all three crewmembers suited to go through where all the hoses are going to be; who s going to plug into what, when, and where; who s going to help who; and what the COMM situation is going to be.

ALDRIN   That s true, but doing that in one g is rather unrealistic.

COLLINS   That may be, but instead of just getting a briefing on where the things are going to be, you ought to see them with your own eyes, particularly the geometry of the thing without the center couch in there, the locations of the hoses, and again, who s going to stand where and who s going to help who plug into which hose when. Three men in there with pressurized suits who don t understand what s happening and can t talk to one another would be one hellacious mess.

24.19 MOCKUPS AND STOWAGE TRAINING EQUIPMENT

ARMSTRONG   The mockups and storage equipment were used extensively and, in general, they were satisfactory for developing the procedures. I think the place where they are short is particularly in the area where you re making connections to and from the spacecraft, operating several different life-support systems such as the EMU, OPS, spacecraft suit loops, and things like this. It s very important that you operate all those valves per equity and know why you are operating them in that manner. Our mockups do not do that. They are just knobs and you just do them. It s important to know why you are doing them.

ALDRIN   It s unfortunate that operating gear weighs so much in one g. It s virtually unbearable to have that piece of hardware on your back for a long period of time.

ARMSTRONG   I think we have a lot of complex flights ahead of us in main-line Apollo and I think there are enough of them, enough possible contingencies, and enough training yet in front of us that it would pay to upgrade this area. Many people still have to learn all that hardware; its whys and wherefores.

24.20 PHOTOGRAPHY AND CAMERA TRAINING EQUIPMENT

COLLINS   I just think that gear ought to be available earlier. It s one of the things you can get done or at least get started on 3 or 4 months before the flight, and yet it s not available. It s another one of those late-arrival categories. I m not sure whether it has to do with the quantity of the training equipment or the fact that we have to get one flight down before we can get around to providing for the next one. I think the familiarization with the cameras (taking them home and taking pictures while you re flying around the country in T38 s) should be done early and not the last couple of weeks. From the flights that I have been associated with, it seems to me that it s always been the last month when that stuff magically appears and they want to talk to you about it and all that; it should be done earlier, I think.

ARMSTRONG   That gear should also include flight fidelity. It should have decals on it; we should be used to seeing the kinds of decals, exposure guides, and things like that that we are going to be using in flight. Those things never show on any of the training equipment. That should be included as mandatory.

ALDRIN   The LM photography is tied to operating in the vehicle in many cases, such as the surface photography with the 16 millimeter attached in various ways to the window bar, to the mirror mount. It looks to me like there is room for significant improvement in this area. I think we got into this a little late in the game. So much of the documentation of a flight depends on the photography. It looks to me like we could use some particular training sessions taking real film with flight cameras and the highest fidelity mockup you can create. I don t know how you would do it really - get the proper lighting conditions.

ARMSTRONG   We had a camera session after we had moved to the Cape, maybe a month before launch or so, where it was quite clear that all the photographic details had still not been incorporated into the flight plan, that is, all the thinking that s involved in planning camera placements, the things you want to take pictures of in the field of view possible in that attitude, and lighting on the subject so that you re getting the details. All that sort of thing needs to be worked out by the photo people rather than the crew. It should be done much earlier in the cycle than it is now. I suspect that when we look at all our films, many of them will show that we suffered from not really understanding exposure or lighting well enough over all situations. That s a weak area.

ALDRIN   I think the photography for power descent, power ascent, that sort of thing, should be worked right into the simulations. You ought to activate the camera in the LMS; take films. You re not going to come up with anything, but you get in the habit of doing this.

COLLINS   Somebody with a fair amount of experience and background should really be concerned that the pictures we bring back are of the correct events, and that they have been properly integrated in the procedures. I am sure that there are probably people over there in the photo lab that are vitally interested in that. Yet they re not in any way in the loop and probably properly so. They are not any way in the flight-planning loop. If we don t feel like taking a picture or something, or if it doesn t occur to us to get it written into the flight plan, it just never happens. I mean there is really nobody who s responsible for the overall photographic excellence or the photographic planning of the flight. If we happen to think about it and if we happen to personally ink it into the flight plan, then it will get done. And if we don t, then it doesn t get done.

ALDRIN   That s the way most of the photographic entries are, pen and ink.

COLLINS   Maybe we didn t spend the amount of time we should have studying that photo plan, but, again, I say if the crew doesn t take a particular interest in it and make sure that it s in there, then somehow it just doesn t get in there.

ARMSTRONG   Maybe there should be a responsible person for each particular flight. Maybe there is, but I can t tell you who it is on our flight. Helmut Knehnel has to run his shop, though.

ARMSTRONG   What I m saying is the project engineer should be over at the lunar-surface simulations; he s up at the CMS, he s over at the flight-planning sessions, and he s trying to integrate all these things to assure that the right film and the right camera s in the right vehicle at the right time and it all plays.

ALDRIN   Just as an example, it doesn t seem to be a very professional approach to the handling of the 16-millimeter camera to tape it with a piece of tape in the focus to the infinity position. If you want to get it to stay in that position, put some screws in there to make it retain that position.

COLLINS   I think it s ridiculous that we don t have some sort of automatic exposure control or automatic light control, or whatever you call it. Commercial cameras are available where all you do is point and click and the thing is in the right exposure value. And there are even cameras available that have switches where you can have either a wide field or an average exposure value to give you a broad coverage. For example, if you took black sky against a white booster, it would average out the black sky and the white booster. It would give you the average value that might not be optimum for either one, but it would be an average panorama. Then if you wanted to be specific and throw a little switch, which zonks a light meter down to a spot-meter kind of thing, you can either point it at the dark sky or point it at the booster. These things exist. It s easy to say, well, you can t qualify them, or the right company doesn t make them, or they re not rugged enough, or they won t pass the salt spray, and otherwise raise barriers. If that had been aggressively pursued, we would have right now in our hands an automatic camera that would take a hell of lot better pictures than we are capable of taking, and we could have qualified the thing by now. I think that should be done, I really do. I think they are pursuing it with Hasselblads, but, my Lord, they have been pursuing it with Hasselblads for years, ever since the subject first came up, and I just don t see any results. Yet we do carry great huge spot meters whose utility is questionable, and we manage to develop and carry those frapping things. That Minolta spot-meter was not used during the flight. I don t know what flights have used it but I d gladly swap it for an automatic light control in a camera. That 2-pound battery is nothing more or less than a handle crank; I d gladly swap it for an automatic light-meter built into the camera. I think we really spend our time and our money going down the wrong road in that camera shop. There may be very real reasons why what I propose is impossible, but from what little I know of it, you ought to have the capability just to point and click and get the right exposure.

ALDRIN   I think the importance of documenting events was extremely well brought out in this particular flight in that we were too busy doing other things to tell exactly where we were in powered descent. The film was able to do this, but I think it did it in a marginal way. I don t believe that the mounting and the field of view that it gets in the right window is anything near what it should be to get documentation of the powered descent and the powered ascent. Another example is the problems that we had in docking. I think that that should have been documented using high-speed motion film from the LM. There is just no way of doing it.

COLLINS   I never used the spot-meter in the command module. Did you ever use the spot-meter?

ALDRIN   I looked through it for some interior settings. I put it on the Earth, but I don t think anything significant was learned.

ARMSTRONG   We decided, based on the spot-meter reading, that we were probably one f-stop off when we got pretty far away. We had to open up one f-stop to f:8 from f:ll.

COLLINS   When was this?

ARMSTRONG   When we were halfway out to the Moon, I guess. That s what your measurements indicated. I never used the spot meter.

ALDRIN   I take it back, I never used it. There s no doubt that an automatic device would be far superior to anything you get out of the spot meter. I take it back, I never used it.

24.21 LUNAR-SURFACE EXPERIMENT TRAINING

ALDRIN   I think we did quite a good job in having fairly high fidelity equipment (solar wind, EASEP package) available. The only improvement I would suggest is that we try and gear it to the type of surface environment. The little problems that we ran into were associated with the interface of that gear to the soil conditions; examples are planting the solar wind and attempting to level the seismometer. This is going to be very true when we start getting into more complex exercises with the ALSEP. To do this on a linoleum floor is almost a waste of time. I think you ve got to do it in realistic conditions by simulating both lighting and surface texture.

ARMSTRONG   It would have probably been better practice to be on a more realistic surface. You probably would have consciously looked into those aspects more than we did on the level surface that we did most of our work on.

24.22 LUNAR
LANDING -
LLTV, LLRF,
LLTV S, AND
LMS

ALDRIN   Like the LEC operation, the big difference that you noted was the effects of the dust getting all over the tapes and cluttering up the cabin.

ARMSTRONG   For the type of trajectory that was required for us to fly (with a long manual flight at the end), the LLTV was a most valuable training experience. Like all simulations, it s primarily a confidence builder to derive the required information from the information that s at hand. In the flight situation, the information that I used in the landing was primarily visual. It was augmented by information inside the cockpit that Buzz relayed to me. I did very little gauge monitoring during the final descent, that is, below 300 feet. It is primarily an out-the-window job, picking a suitable landing spot and getting into it. The full-scale simulations are the only ones that do this - the LLTV and the LLRF. I would have to recommend continuing them both, at least until we have a few more landings under our belt. I would suggest that more attention be given in the LLTV to changing your landing spot while you re in the trajectory.

ALDRIN   And how to deviate from an automatic trajectory and smoothly pick up what you want to do in the way of deviations.

ARMSTRONG   I believe the LLTV can do that job and do it safely. That means that you probably have to do a few more total trajectories than we did in preparation for this flight. I suggest that a dozen is a desirable number - a dozen lunar trajectories in the LLTV. It takes about half a dozen before you re comfortably flying on a lunar trajectory, and after that, a couple of different deviations to different touchdown areas. The LLRF lighting simulation was quite interesting, but in retrospect, it s not a very good simulation of the lunar lighting situation. In the flight, you see much more daylight, at least at our Sun angle (10-degree Sun angle). It was much more of a daylight landing situation than the simulation that was portrayed by the night lighting simulation at Langley Research Center.

ALDRIN   They essentially set up a situation where there was no available horizon. That certainly was available in the actual case.

ARMSTRONG   The LMS new model is really a fine addition to the simulator. If you could afford building a model for Apollo 12, so that their last 2 months of simulation would be going into the Surveyor site, then I think you would get a substantial improvement in your confidence level to get to the desired touchdown site.

ALDRIN   I think this is particularly true if they stick to the objective of going to that specific area. We have enough available information from the Surveyor itself to build that model.

ARMSTRONG   I know that s an expensive item to provide, but our experience with looking at the L&A of Site 3 indicates that you really can get a good understanding of that local area in your many landing simulations in the LMS.

ALDRIN   In looking back on the choices that I made with regard to my participation in landing simulations, I think they were generally correct. I don t think that I suffered by not being exposed any more to the LLTV. I think one session at Langley was worth the effort. I concentrated on manual use of the throttle and I think that s probably what future LMP s should concentrate on, also. I think Neil agrees that if we did have to execute a complete manual landing, it would probably best be done by the Commander concentrating on attitude control and voicing to the LMP what rate of descent and what changes he wanted. It appeared to be a very difficult task for one person to accomplish all of these. Whereas, when the tasks were split, and use was made of the instruments to manually control the throttle, and a fair amount of practice was made, use of that good performance could be anticipated by a manual throttle landing. For the most part, this can be done in the LMS.

COLLINS   As a general comment, our support was inversely proportional to the number of days remaining before the flight. We had poor support at first and later we had superlative support. I would have traded some of that last-minute support for some earlier support. To be more specific, early in the game, the flight-planning people, and the checklist people, and the command-module rendezvous procedures people worked for three different bosses and lived in three different worlds. It was not until late in the game that John O Neill was given the overall power and you could go to John or somebody he designated and say, Look I ve got this problem. The checklist says one thing and the flight plan says something else and the North Americans have never heard of either one of them. That would get squared away. But early in the game, it seems to me that the checklist people sort of pointed the finger at the flight-planning people who responded by pointing fingers in return, and a lot of time was spent, you know, looking for a left-handed monkey wrench. You sort of wandered up and down the second floor of building 4 trying to find somebody who would really take the time and be interested in researching the problem and coming up with a procedure technique. Late in the game, it was all amalgamated under John O Neill and it worked as I think it should have worked from the beginning. I don t understand the breakdown in FCOD; I don t understand Ernie Dement s bailiwick as opposed to the flight-planning world. It seemed that much of the time, they were working at cross-purposes and it seemed like - I guess they get negative vibrations from the checklist world. I guess I have to say that I don t understand their problems fully and perhaps they don t understand my problems, but I don t enjoy making changes to procedures. It seems like the crew only does that when they feel there s some good need for it. And yet the checklist people seem to have the feeling that other flights have gotten by with this procedure and why can t you.

ALDRIN   It seems to me that they were unwilling to meet us halfway. We had a different job to do, different hardware changes in the spacecraft, and different uses to make of the equipment.

COLLINS   Maybe it s an unfair comment, but I had the idea that their viewpoint was that it was good enough for previous flights and it s good enough for this flight, so don t bug us with changes.

ALDRIN   I agree with what you re saying, Mike. I think it s unfortunate that there was a division between the checklist people, the procedures-development people, and the onboard data. It seems to me that the procedures-development people should be the ones who also work with the handbook. They should start it and carry right on through completion, which includes onboard data. The sooner the crew can start training with data books that represent the best of your ability at that stage of training, the more they are going to get out of it. We had several new areas, and it appeared as though we were pioneering much of this in the areas of procedures development, and also in determining just how this was going to be presented to the crew and how you make use of it, and in distributing it around the spacecraft. We had to make certain decisions, and we tried one form and then another. I m sure there are better ways of doing it than what we settled on, had many things been done before we got on the scene. I sure hope that follow-on crews won t find it necessary to make big changes.

ARMSTRONG   We had five straight flights here on very close centers. Each crew has been obliged to get some procedures that work and stick with them, to settle on them close enough before the flight so that they could remember what they were. That meant to each and every flight, I m sure, that there wasn t time to sit around and figure out which of several different approaches was the best. We had to take one that worked and stick with it, and in many cases, this resulted in choosing one that clearly wasn t the best, but it was one that worked. The next flight that came along was obliged, wherever possible, to take everything that the previous flight had been able to work out and to go with that. They had enough of their own new things to be concerned with; they had to choose one early that worked and go with it, and not spend too much time deciding which one was best. At the end of five flights here, the result is, I think, that we have the procedures for a lunar mission, about 60 percent of which are not the best ones to use. They are ones that work, but they are a long way from being ideal. We have a little more time before the next flight, and I hope that during that period, we can take the ones that are good and do that work and use those, but not hesitate to change those that are really marginal procedures. That s going to take everybody s cooperation to pick out the marginal procedures and to improve those to the level you would like to operate with for the rest of the lunar program. It was an inevitable conclusion of the schedule that we were forced to meet. Everybody had his nose to the grindstone to make the thing work. Now we just have to accept the fact that the inevitable consequences of that situation are that we don t, in fact, have the best of everything at this point. This tends to be a lot of adverse comment, and it really shouldn t reflect that, because the facts are that when you look at it in the overall sense, it did the job. It got us ready to fly and, essentially, we didn t have any big open areas. In the overall sense, it is damn good. I think we tend to be very self-critical in this area, though, because we ve all had our hearts and souls in it for a year or so.

COLLINS   I try to put it in perspective myself and say that I thought it was, all in all, an excellent training cycle and very good use was made of our time. We had wonderful support. In some cases, that support came very late, but we had, I think, beautiful support - and I thought it was extremely well worked out, considering the complexity of the things we had to learn. I think that, just from the CMP viewpoint, the proficiency of the CMP (all other things being equal and they probably aren t) is just proportional to how much CMS time he gets. I thought that I was adequately trained but that I really wasn t particularly polished in any one area. I just didn t have the time to devote to each and every little slice of the pie. I tried to learn all the systems; I tried to learn all the procedures for burns, all the rendezvous procedures, and the navigation, but I will be the first to admit I was far from being an expert in any one of these fields. I don t see how you really can be an expert unless you have more time, more simulator time, to devote to it than I had. I think 400 hours should be a minimum. The only reason I bring this up is that I think some of our training plans say something around 200, 250 hours is sufficient. I don t really think that s true. Speaking from the command-module viewpoint, I don t think you ought to be launching CMP s with less than 400 hours of simulator time. I really don t.

SLAYTO N   A lot of your time was spent developing procedures.

COLLINS   There s a lot of truth in that; there were some areas that I had to work but that had not been worked out before. Even if you deleted all those, however, I still think a lot more than 200, 250 hours is required. You just take the pie and start slicing up the lunar mission; you take all those systems; all the malfunctions; the various mission phases; and the if s, and s, and but s ; of the various rendezvous. I don t think you could cram that into 200, 250 hours.

ARMSTRO N G   I would guess that if you would look at the integrated simulations with MCC and total up the hours there, I would bet it s a significantly larger number than we used to think about per flight. It s because of the many phases of this complex mission; there are just so many phases and each has to be covered in fairly large amounts of time. It s good time, but you really can t count it towards your basic training for the mission.

COLLINS   You should go into those integrated simulations having all the basics behind you; that s just sort of the graduation exercise in a particular phase of the flight. I just wanted to mention a minor point. I did fly a couple of entry sessions on the FOD s entry simulator, which is an awful-looking little thing on the third floor over in building 30 with a bunch of old Gemini components and make believe DSKY s. However, it comes with John Harpold, who understands the entry math flow probably better than anybody else that I know around here. I think it is worthwhile to schedule just as I did maybe two 2-hour sessions on that thing, and I don t understand why the CMS cannot do this. Harpold can crank failures, accelerometers, stuck accelerometers out of SPEC, and little internal failures into that simulator and show you how the computer would handle them - in most cases, how it would fail to handle them - and he has failure modes in his simulator that, so far at least, they have been unable to crank into the CMS. I consider that a worthwhile exercise. It would probably be even better to incorporate those failures somehow into the CMS - he has tried to do it but has been unable to do so. I am not sure whether it is the limitations of the interpretator or what it is, but I think that training was worthwhile.

ALDRIN   We haven t covered one category about suited operations. We did a fair amount of suited training in both simulators, and I am certainly glad that we had that amount of time. I can t really identify many areas where suited operations in flight proved to be a big hindrance, but I think the sooner you can begin to integrate the total mission package with the data, the type of pens and pencils you are going to be using, where you are going to put them, and where you log all the data under suited conditions, means a higher fidelity training, and I was glad that we did as much suited operations as we did.

ARMSTRONG   Did you keep track of our suited operations overall? We spent much more time in those suits than I ever thought anybody could spend in preparation for one flight. Almost every day for 3 months before the flight, we were in that suit sometime during the day. It would be nice if you didn t have to spend that much time in the suit, and perhaps we didn t, but I guess we had a high degree of confidence in our ability to operate in the suits in the various environments. I think we probably spent more time in the suits than we had to. We did spend much more time in EMU CCFF, fit checks, and stuff like that than we should ever have had to. Seems like we did about 10 of those exercises, and every time they would change a little something on the underwear or something, they would want another fit check. We bowed to most of those and did them, but I would hope that future flights won t have that much instability in their configuration.

ARMSTRONG   I think we have a tendency to reflect on what we did and to say that we made the right decision there. I m not sure that we re in a good position to really judge. There are several simulation areas that should be improved, some of which we ve mentioned before. There is one other area I don t think we discussed, that of optics, and I think it s true in both vehicles, certainly in the LM. The AOT optics characteristic limitations and constraints, such as lighting and sun shafting here and there, are not covered at all in any of our simulations, nor at any time do we actually get a very good opportunity to look through real optics and understand their limitations. I really think that we need some optics someplace that look at the real sky, the real constraints, the real illuminations, side lighting into the optics and things like that so that we can appreciate what you can and can t see. I m not even sure that anybody agrees with me in this area.

ALDRIN   I agree with that. I think that for the surface alignment, I was quite surprised to find that four out of the six detents were unusable for the surface alignment. I wouldn t have thought that beforehand

ARMSTRONG   They become very significant on some flights when you start talking about particular little details of the flight. Our mission simulators just don t cut the mustard in this area; they re going to tell you answers that are wrong. I can understand; I just think it s an inherent limitation the way those simulators are built, and I think we need to augment that somehow with some real optics with real lighting. I m not quite sure what the best way to do that is, but I think you could do it with existing hardware, prototype or test hardware. You could get some of that stuff together and build a special simulation that would give people the opportunity to work, before launch, with some real optics and to mark on some real stars or something.

COLLINS We ve done that at MIT to a very slight degree. From the command module viewpoint, I d have to say that that sure would be nice to have but I can t think of any situation where a lack of that high fidelity training would make you come to a dangerous conclusion - maybe a wrong conclusion or maybe you might get tricked into thinking you could see the LM farther away than you really could or that you could see more star patterns than you really could.

ALDRIN I think if we had to do P51 s where your initialized identification would have to be done by the telescope, it would have pointed out many of these deficiencies.

COLLINS That s right. I thought about saying that they ought to change the characteristics of the telescope in the CMS but I think if you had to do a P51, what you d do is turn out all the lights inside the command module and, if necessary, put a bag over your head and take the 20 or 30 minutes to dark adapt. Now, you can tweak the CMS telescope to that same level but then all you ve really done is wasted a lot of simulator time, because that means every time you look out through the telescope, you have to wait 20 minutes before you can see anything. That s really all it means. So I don t know. I think in regard to the telescope part of it, the simulator should be left like it is. Concerning visual presentation of the LM as a little pinpoint of light in the sextant, for example, during the rendezvous sequence, that is unreal, but I m damned if I know how you make that real. I just don t know how you d do that. Maybe it s doable but that LM as it goes off against the lunar surface background gets smaller and smaller. It is one problem at 50 miles, a different problem at 100 miles, and a little different problem at 120 miles. To have high resolution optics

of that type seems to me to be beyond our capability. I don t know how you d do it.

25.0 HUMAN FACTORS

25.1 PREFLIGHT

ARMSTRONG   I guess our only activities that fit in preventive medical; procedures were disinviting the President and trying to slow down to a reasonable or, at least, acceptable pace in the last week or so. The Cape doctors kept an eye on us that last couple of weeks and I guess we didn t have any complaints there. They did a good job. We got through medical briefing.

ALDRIN   I had a couple of conditions come up in the category of dental care and I thought there d be plenty of time to get them taken care of but there were things that came up at the last minute and we were hard pressed to schedule those in. I would highly recommend that people really take a real close look at their own status as far as those things go and get those things taken care of as early as possible.

25.2 FOOD AND WATER

ARMSTRONG   Comparing hunger sensations in-flight versus two weeks preflight, I d just say, in general, that I didn t have as large an appetite in-flight as I would on the ground, but I thought it was adequate and I was able to eat enough.

ARMSTRONG   The food was palatable and all three of us kept our levels up satisfactorily, I think.

COLLINS   My appetite was off on the first 2 or 3 days of the flight, I would say. After that, it was close, if not equal, to my usual ravenous ground appetite.

ALDRIN   I didn t find any difficulty in generating a desire to eat.

ARMSTRONG   I agree with that.

ALDRIN   I think (laughing) in comparison with Gemini, it was good. There were times in Gemini when, of course, we didn t have enough time to do this because food preparation is a very time-consuming task. During the translunar and trans-earth coasts, there s plenty of time to take care of it; but, no kidding, it takes a long time to get these things ready. If you do have a lack of appetite, the tendency is just to forget about doing a lot of that stuff. But I didn t experience a lack of appetite at all on this flight.

ARMSTRONG   Comments on the taste.

COLLINS   In comparing food during preflight evaluation and in-flight taste, I noticed no difference.

ARMSTRONG   Acceptability of the foods. Well, just first make an overall comment that the new foods are significantly improved and welcome additions to the menu. I think, in general, it s a real aid to the normal day-to-day operations in the spacecraft to have pleasant menus and palatable meals.

ALDRIN   I think that most people are aware that during translunar coast we did, for the most part, make use of the prepackaged meals. I guess partially because we knew they were more of a low residue and we wanted to avoid any complications with waste elimination that might interfere with the LM activities.

COLLINS   In general, I thought the food was at least excellent or better. I thought a lot of hard work went in on the food selection. In general, I thought the quality of the food was extremely good. My criticism of the food revolves around the packaging. I think we waste too much time fixing it; and, for this particular flight, there was more food than three people could have eaten in 3 weeks. They really gave us a lot of food. I think they probably don t need to provide nearly as much. We had our normal three packages of food, plus this little pantry arrangement which is very convenient and nice, plus we had a bunch of wet packs. I d say we probably ate half the food onboard, we had good appetites and we ate - I d hate to say how many calories per day, but plenty per day.

ALDRIN   The one disappointing package, I guess, was in the wet packs. The turkey and gravy I thought was outstanding because it was moist. That wasn t the case with the ham and potatoes, nor the beef and potatoes. I thought that both of those were too dry and that the potatoes were not appetizing at all.

COLLINS   In general, I found that the sweet things were not as good as the others. This applied to the drinks as well as desserts. I touched very little in the way of desserts. On the drinks, I felt that something tart, maybe like limeade, would have been a nice addition - or iced tea or something like that.

ARMSTRONG   I agree. In general, I felt the beverages were too sweet.

ALDRIN   I think that we can go still further than we have in the line of the canned spreads going on either bread or toast. I just didn t experience any difficulty at all in zero g taking a spoonful of this and spreading it out. As long as the material that you re using is relatively moist, it stays together. It doesn t have a tendency to run off and go all over the cockpit. We had tube spread in the LM, and I think we could have used that type of a preparation in the command module, along with more of the canned variety. Of course, the canned variety presents a problem of disposal afterwards. It ll certainly have to be reckoned with. I m not sure how you make use of a pill or disinfectant with cans.

ARMSTRONG   The spoon-bowl items were fine; intermediate moisture fruits, sandwich spreads, and breads were all used extensively. In general, I liked the pantry approach. I thought the approach where you went in and selected those items that you would enjoy for that meal and assembled your own menu was a very pleasant operating mode. I enjoyed that, if you could handle your diet satisfactorily that way.

ALDRIN   I think it would be a good idea to package the pills along with the spoon-fed packages, either that, or have some different, more convenient way of dispensing them. After meals, you gradually dispose of things as you re consuming them, and you don t want to have to get up at that point and float down to the pantry to get the pills out to pop one of them in the bag.

COLLINS   I don t know. The business of disposing of all the waste packaging, putting pills in, and all that is very time consuming and creates a huge volume of waste. Really, the way to do it is to use the pill, then wad everything together, tie the little packet as tightly as you possibly can, and wrap some tape around so that it stays in a small volume, high-density package; however, this is time consuming. It would really be nice if you could have some thing like a commercial garbage disposal where you could just take all this stuff and cram it in, turn a crank, pull a switch, and have it all sort of ground up and disinfected and spit into a stowage compartment.

ALDRIN   Either that or some sort of an airlock where you could take this refuse and put it in the airlock and dump it overboard.

COLLINS   Right. But the packaging, getting the food reconstituted, and then doing something with the other packages were the biggest drawbacks. Breakfast would have been improved, I thought, if they had some scrambled eggs, which I know they have in the lab. I just don t think they have gotten around to putting them on the flight menu, but it would sure be a good idea if they did have some of that. It would be a welcome addition to breakfast.

ALDRIN   In the right-hand seat, I found it convenient to take some of the Velcro that was on the food packs and put it on the scissors. I just found that with the scissors at the end, the cord was just a little bit too unwieldy. That brings up another point. When you put those scissors and things like that in the pockets where you have the dosimeter and a few other things the pocket just doesn t seal right. You move around a little bit and pretty soon you re missing a flashlight, you re missing a pair of scissors, and the dosimeter is off somewhere.

COLLINS   That s exactly right. Now, these in-flight coveralls are carefully tailored garments and a lot of engineering has gone into them and they are almost half as good as the summer flying suit. You don t have the problems like that with the summer flying suit because they have zippers in all pockets and you are accustomed to using them.

ALDRIN   Do you care to say a few words about snaps, Neil?

ARMSTRONG   There ought to be a law against snaps. I think that if I were preparing for this activity, for further flight, I would take just a piece of cloth and sew a bunch of little pockets and separators in, and then have a place for all the little odds and ends that you like to keep handy, like your scissors, tooth brush, spoon, pencil, and a bunch of things like that - keep it in one pocket.

ALDRIN   One for each individual and a reasonably convenient place for each crew station to mount that sort of thing. There is one now in the LEB and it s a little bit too large.

ARMSTRONG   That would have helped keep track of all those little loose items that are just personal necessities. As recorded on previous tapes, we were periodically losing some piece of equipment, a toothbrush would be gone; a camera back, a monocular, or some tape recorder would be drifting around the spacecraft somewhere and it would be a matter of going on a big search to find it.

COLLINS   The area behind or above your head in the left-hand couch and the right-hand couch is a convenient area, because it is an uninterrupted bulkhead space very sparsely covered by little patches of Velcro; that was the place where we wanted our cameras, monocular, and tape recorders. If there were a couple of spring clips up there, built to be the width of the Hasselblad or if there were more Velcro in that area, it would be a lot more habitable. The doggone camera was always floating around, because there was never enough Velcro really to plaster it up against the wall.

ALDRIN   How about a comment on the coffee?

COLLINS   The coffee was a little disappointing; I don t know what was wrong with that coffee. It wasn t a good brand or the coffee was very tasteless - not tasteless, but it just had a peculiar taste. It didn t taste like coffee. I like instant coffee; I drink it at home all the time.

ALDRIN   It was pretty hot.

COLLINS   The water was hot; now I don t know what it was, but the coffee just wasn t very good.

ARMSTRONG   In general, the water wasn t that good. There was a little chlorine taste to it. I found that I drank a lot of cold water instead of the other beverages that were available and I enjoyed it. There was some gas in it. There are some engineering improvements that can be made in the filters, but I think that s possible.

COLLINS It s well worth carrying those filters. You need them. You might put on the tape that it would be nice to have a fingernail clipper onboard the spacecraft. I don t know where the best place for it is, but it would be very convenient to have in case you rip off your fingernail or get hangnails. At present, there is no tool available that will reach that area.

25.3 WORK-REST CYCLE

ARMSTRONG I guess overall, with a few exceptions that were just discussed, the work-rest cycle on this particular flight was reasonably good. We were essentially operating on Houston time. We were getting our simultaneous sleep periods and essentially it was during the sleep period here in Houston.

ALDRIN All of us elected to have Houston time on our watches and I think it was unfortunate that we didn t have the flight plan also geared to Central Daylight instead of Eastern Daylight. At the top of each page, we had the corresponding time for Eastern. It would have been an improvement, I think, if that had been Central.

COLLINS Yes, I think they figured that the crew was on Cape time so print local Cape time; but, really, we were on Houston time in our minds. It s a small point.

ALDRIN I m sure the Control Center would have preferred it the other way too.

COLLINS Sure.

ALDRIN I had anticipated considerably more difficulty with getting adequate rest, especially the first day. But it didn t turn out that there was much of any problem at all. I thought the sleep stations were very comfortable and the temperatures seemed to me to be very pleasant. I think coming back we noticed that it was getting a little cooler.

ARMSTRONG It was a little warm in the daytime. It was a little cool, particularly at night, on the way back.

COLLINS I think it is important somehow on these lunar flights to get yourself in the frame of mind where you regard the first couple of days of flight as just preliminary to the lunar activities and somehow you talk yourself into relaxing, taking things easy, and getting adequate sleep the first 2 or 3 nights so that you don t arrive at the Moon already tired when the peak activities begin.

COLLINS Maybe this is belaboring the obvious, and maybe all crews know this and will think about it, but this is something that we talked about; I think it is kind of a frame-of-mind thing. I think you can talk yourself into either getting all excited and burn up a lot of energy in anticipation or, on the other hand, you can talk yourself into relaxing and taking things easy. Personally, I felt that having flown once before was very helpful to me. I had been up there in zero g before and I wasn t spending all my time pondering the wonder of it all. I was in a familiar place and I was willing to pretend the flight hadn t started until along about the time of LM separation. I think this is important for these flights with extended lunar-stay times, particularly when the crew is flying a flight for the first time. Somehow they ought to talk themselves into a proper frame of mind and get good sleep and arrive at the Moon in a rested condition.

25.4 EXERCISE

ARMSTRONG We all did a little bit of exercise almost every day. We used either isometrics or calisthenics in place of the Exer-genie. The Exer-genie worked alright. It got a little hot and stored a lot of heat, but it was acceptable

COLLINS   If you got a good workout on the Exer-genie, it got so hot that you couldn t really touch it. I don t think that is any kind of problem; I just mentioned it.

ARMSTRONG   Any other comments on exercise?

COLLINS   I had the idea that it was worth exercising on the way home and maybe not worth exercising on the way out.

ALDRIN   During the lunar-surface activities, it didn t appear to me that preconditioning in any extensive degree was required. Now, if you were going to take 7 days to get there, it might be a different story. Certainly, with the activities that you have in one g, you are not going to deteriorate that appreciably in 3 days.

COLLINS   Well, I felt better in the water when I was first back in one g and stood up in the lower equipment bay. I felt a lot better on this flight than I did on the Gemini flight. I am not sure what to attribute that to. If I had to guess, I would say maybe having the suit on in Gemini and having it off on Apollo - having already stored a lot of heat when I arrived at that point on Gemini and being cool and comfortable on Apollo; maybe it had something to do with exercise or the increased volume inside the spacecraft - I don t know. But I felt a lot better and I felt in much better shape this flight than I did on the Gemini flight.

ALDRIN   It goes back to what you said before. I think the fact that you have been there and have been exposed to a landing on the water and seas that are not calm as can be - I think having been through it once - the second time does make it a good bit easier.

COLLINS   Maybe that s it. But I can remember a heaviness in the legs on Gemini. I could just visualize those legs being pooled with blood. It seems like the old heart just wasn t capable of pumping things uphill as it usually was. I felt heavy in the legs, and sort of foggy, and I didn t feel very good. This flight, I didn t notice it at all.

ALDRIN   I think the difference in the space available inside the cockpit enabled you to move in a fairly regular sense and that just wasn t true in Gemini, where you were sitting and didn t get the opportunity to stretch your legs out.

COLLINS   I couldn t stretch all the way out in Gemini. My head hit, or my feet hit first.

ARMSTRONG   Generally, I can say that we didn t have any problem there. The toothbrushes and toothpaste worked fine. Essentially, we followed the normal pattern just as we would on the ground. As a matter of fact, not in just that area, but in as many areas as we could - eating, sleeping, normal habits, workdays, and so on. We tried to follow a normal pattern as we would on the surface. I think that contributed to the fact that we felt good the whole time, felt rested, and were able to do a good job.

ALDRIN   I thought the toothpaste was pleasant.

COLLINS   Yes. I brushed my teeth twice a day and everything was normal in that regard.

ALDRIN   In the category of sunglasses, I found that they were of considerable use in Gemini; however, in the command module, I didn t see any use for them at all. Now, they may have been of more use in lunar orbit. In the LM, there were times when we had our helmets on most of the time.

ARMSTRONG   I used the sunglasses for a while, early in the flight, and then chose not to use them anymore.

COLLINS   Yes, I think they are questionable. I would not suggest deleting them. Some people use sunglasses extensively. I know some people, whenever they go outdoors, clamp sun glasses on their eyes and maybe those people would do the same thing in spacecraft. I don t use them very much on the ground. I only use them when driving a car, but other than that, I rarely use them. I would like to go back to light attenuation under sun glasses. On Gemini, we had a window shade with a Polaroid circular filter in it. I thought it was a little jewel. I tried to get that added one time to Apollo and the CCB turned it down. In retrospect, I certainly couldn t say that that s something that you absolutely have to have, but that would surely be nice for window number 2. I don t know if you all remember; but, during rendezvous, you have to look at something bright. It s great because you have this little circular Polaroid section that you just rotate to any angle you want to get any degree of light you want. It would be a very useful addition, I think, to this storage list. Now, I can t say that it s necessary or mandatory, but in a nice-to have category. I d sure swap my sunglasses for that light attenuator any day. The reason is that, when you put the sunglasses on, you not only attenuate the outside light (which is desirable) but also attenuate the inside light, which is undesirable. With the screen on the window, you can filter as much light as you want and still read all your gauges with complete clarity. If you put sunglasses on to block the outside view, it also blocks your inside view. So I guess that s my little speech in favor of that sort of light filter. It would be nice to have. Speaking of window shades, this may not be the best place to bring it up under human factors, but I like to have my sleeping accommodations dark, as dark as I can get them. Certainly the window covers were good, but they weren t as good as they could have been. They were quite difficult to install and I don t know what the reason for that was. We fit-checked them and I don t recall any difficulty fit-checking them on the ground. They were very tight, but not nearly as tight as they were on flight. We got more exercise wrestling with the window shades than we did out of the exerciser, I believe. Every night, we had a 10-minute yell-and scream-and-swear session, and jump up and down trying to force the window shades into place.

ALDRIN   And they still ended up leaking a certain amount.

ARMSTRONG   It looks like it might be advantageous to have a cinch down mechanism on those window shades that had a higher mechanical advantage than the ones that are on there now.

ALDRIN   Yes.

ARMSTRONG   The ones that are on there now require a tremendous amount of force to engage.

COLLINS   They require a tremendous amount of force to jockey into position where the lever would fit over the top of them. Then they require an awful lot of additional force to get the lever over center.

ALDRIN   It might be interesting to note at this point that, in regard to the spacecraft lighting, I think while we were all asleep was the only time that we really made use of the back lighting and the EL lighting. Maybe this is enough reason for it to have it readily available, so that you don t have to flick on the floodlights, but other than that use, I don t think it s required.

COLLINS   The EL light?

ALDRIN   Yes.

ALDRIN   Very nice, very pleasant to look at, but we just didn t need it, I thought.

COLLINS   Block I used to have floodlighting alone. You d have to look at a vehicle with floodlighting alone under a lot of different circumstances. If I remember, Block I used to have shadow areas where the struts would get in the way between the light source and the gauge and things like that. I d sure hate to go back to that kind of a lighting scheme. It s nice to have the EL.

25.7 UNUSUAL OR UNEXPECTED VISUAL PHENOMENA

ARMSTRONG   Okay, visual phenomena have already been discussed.

25.8 MEDICAL KITS

ARMSTRONG   One comment here is that it was pretty clear that the medical kits were not carefully packed. The pill containers blew up as if they had been packed at atmospheric pressure. The entire box was overstuffed and swollen. It was almost impossible to get it out of the medical kit container.

COLLINS   I ripped the handle off as a matter of fact, trying to pull it out.

ARMSTRONG   That was even after we cut one side off the medical kit so it would be less bulky so we would be able to put it in the slot. I think that s just evidence of less than the required amount of precaution in packing.

ALDRIN   I guess we have never really covered the distasteful area of bowel movements.

COLLINS   Why don t you cover that?

ALDRIN   Mike sort of indicated that we probably should discuss this area further and there may be some better way of handling waste material than with the bags. It certainly is messy and it s distasteful for everybody involved to do it in that particular fashion.

25.9 HOUSEKEEPING

ARMSTRONG   In general, it s a continual load. There are always things to be done, equipment to be stowed, windows to be cleaned, air filters to be cleaned. There is a continual never-ending bunch of chores to be done, which is desirable in some ways, I guess. It keeps you busy on the translunar and trans-earth coasts. A lot of those areas are required just because of the approach taken toward that particular design, as a lot of the construction in the cockpit - all the stowage equipment - gets put together and assembled in Erector Set fashion. That takes a lot of time and leaves a lot of stuff out. In general, I think it s an area that can still use a lot of improvement.

ALDRIN   I think the idea of having an individual kit where you can place things in individual packages is much better than that large one. And I d like to see continued effort along this line to come up with better ways of interim stowage.

ARMSTRONG   We used the new stowage that was devised after Apollo 10 and it worked okay. There is probably more equipment available there than you really need, but it worked.

25.10 SHAVING

ARMSTRONG   We did shaving onboard and didn t have a lot of real good luck with that. For some reason or other, we let our whiskers get pretty long before we tried that and found out it was an hour s job to shave.

ALDRIN   It takes a lot more water than you d think ahead of time, and getting water on your face is not too easy a task. You can get some to accumulate on your fingers in a thin film and then get it on your face, but invariably it s going to start bubbling and get all over the cockpit in various places.

ARMSTRONG   The only difficulty really was conditioning the beard for shaving. Handling the equipment was no problem and there was no problem with shaving cream getting away from you. It wasn t that kind of a problem.

ALDRIN   Well, it did use up a fair number of tissues to keep wiping it off.

COLLINS   Now, in one g, what you do when you get all through shaving is to bend over the bowl, you take water, wipe it all over your face, and all the bits and pieces of hair go down the sink. But the way we were doing it, when you got through, they were all over your face; then you had to wipe each and every one off. It was sort of hard to get them off. For hours afterwards, they were - scratching and itching.

ALDRIN   I think if I had it to do over again, I would have shaved once before the lunar operations.

ARMSTRONG   Yes, I think it s better to shave more often.

ALDRIN   It was a little bothersome putting that COMM carrier back on again and having a chinstrap going across underneath.

26.0 MISCELLANEOUS

ARMSTRONG   We had some and I m not quite sure why.

COLLINS   What do you mean?

ARMSTRONG   Well, we had to keep the sensors on and try to get data.

COLLINS   I don t see any requirement for that in the command module. I really don t see any requirement for sensors at all. You just have a bunch of extra claptrap, complexity, and power drain.

ARMSTRONG   In general, I think each person should feel like he understands his reaction to various kinds of medications that might be required before flight. I think it s unfortunate when you have to do that kind of stuff in the last week or two before the flight. That should all be done very early in the training cycle so that s no problem or concern to you at that late date.

COLLINS   I agree. Another thing is that, it appears to me, we should have pills in two categories - those that we can take without obtaining permission, and those whose use requires in-flight permission. And I d put the motion sickness pills in the former category. In other words, the point where the motion sickness is. According to the doctors, at the first onset of any symptoms, you should be taking pills. If that s true, I think you should just go ahead and take one without having to go through the rain dance of calling the ground, getting permission, and then having a big conversation go out to the world about how you are; in fact, are you going to throw up or not? If the pills are safe, then I think we should be given credit for having the judgment to decide whether to take one or not.

ARMSTRONG   I don t guess we had any comment there. We thought our preflight requirements shortly before launch were excessive.

ALDRIN   I think we all feel about the same way about having that press conference conducted here through the glass. It would have been far better to have done that sort of thing earlier. I would think that that late in the training is too late to be conducting that sort of press activity. I don t think we created a good impression, particularly by the way we went about the protection.

COLLINS   I felt like all three of our press conferences were bombs. I really did. I guess we have nobody to blame for that except ourselves. I just felt like they were dull and boring, and that very little useful information was interchanged.

ALDRIN   I agree with that. I think that the TV in-flight should be something that depends on the crew s desires. I think the more successful ones that we had were ones that were spontaneous, where we just started showing things around. Then the ground, I guess under the supervision of people in our office, can monitor what is received and then release what is appropriate rather than having TV scheduled at certain times and going out live. I prefer not to handle it that way.

ALDRIN   Put wheels on the thing, so you don t have all these problems of people pulling it around from one position on the carrier to another. I guess they re in the process of thinking about this anyway.

ARMSTRONG   We recorded a lot of comments on the MQF design while we were there. Basically, the operation of that piece of equipment was satisfactory with few people at hand.

ALDRIN   We noted that the table was awkward in its location. It was awkward to move around and took up a lot of additional space. The windows need to be enlarged if they re going to be used for that type of public-affairs activities. I guess the communications to the outside were relatively good.

<u>26.4 LRL
OPERATIONS</u>

COLLINS   I want out.

ARMSTRONG   I guess we don t have any comment there. So far, they ve been going as well as you can expect.

27.0 CONCLUDING COMMENTS

NONE

APPENDIX
APOLLO GUIDANCE COMPUTER

The Apollo 11 technical debriefing makes many references to various programming commands used during the course of the flight. While editing the document it became increasingly clear that some kind of explanation of these acronyms would be necessary.

The Apollo Guidance Computer was a surprisingly simple and elegant system. Inputs to the flight computer were made through the Display & Keyboard (DSKY). This interface was a keyboard with 19 keys, a series of ten warning lights, three five-digit numeric displays, and displays for PROGRAM, VERB and NOUN entries. The Command Module used two of these interfaces, one on the main control panel in front of the commander and the other in the Lower Equipment Bay (LEB) at the navigator s station. A similar DSKY existed in the center of the Lunar Module directly above the EVA hatch. The main difference on the LM DSKY was that it featured an additional three warning lights which monitored altitude and velocity and the digital autopilot. The main DSKY warning lights could alert the user to everything from temperature problems to keyboard errors.

To make any sense of the protocols involved it is necessary to have a basic understanding of the grammar that was used.

In order to have the AGC carry out an instruction the pilot would have to punch in a sequence of codes. These codes were known as VERBS and NOUNS.

When the pilot pressed the VERB key this indicated to the computer that it is going to take some action. The following two digits would then be interpreted by the computer as the VERB code to take that action. An example would be VERB 49 which would tell the computer that this was the beginning of a crew defined maneuver. In the case of the LM an example would be VERB 70 which told the computer that the pilot was going to update the lift-off time.

When the pilot pressed the NOUN key this conditioned the computer to interpret the next two numerical characters as to what type of action is applied to the verb code. For example VERB 6 NOUN 20 would tell the computer that the pilot wants the computer to display the ICDU angles in decimal. NOUN 20 says I WANT TO SEE THE ANGLES and VERB 6 says DISPLAY IN DECIMAL ON ONE OF THE DISPLAYS .

As you can see the choice of VERB and NOUN is quite appropriate. VERB is the action and NOUN is the substantive to which the action is applied.

By using this system a flight to the moon and back could be accomplished with a series of pre-determined two digit codes and something in the order of 10,500 key punches.

While there were approximately 100 VERBS and 100 NOUNS there were also some 70 predefined PROGRAMS.

To enter a PROGRAM in the DSKY one example would be VERB 37 (this told the computer that the pilot wished to change the preset PROGRAM) followed by 23 ENTER. PROGRAM 23 was a Star/Navigation Measurement. Thus when Michael Collins says he did a P23 it means he took a Navigation Measurement using the computer.

Attempting to clarify the meanings of the various codes is an arduous task as many of them changed as the two Apollo spacecraft evolved. Thanks go to Andy Chaikin and Frank Sietzen for their assistance in exhuming some of this information. It should be noted that neither the following Acronym guide or AGC codes were part of the original Technical Debriefing transcript.

Robert Godwin

ACRONYMS

AGS	Abort Guidance Subsystem
ALSEP	Apollo Lunar Surface Experiments Package
AOH	Apollo Operation Handbook
AOS	Acquisition Of Signal
AOT	Alignment Optical Telescope
APS	Auxiliary Propulsion System
ARS	Atmosphere Revitalization System
BCMP	Back-up Command Module Pilot
BIG	Biological Isolation Garment
BPC	Boost Protective Cover
CAL	Calibration
CAP COMM	Capsule Communicator at Mission Control
CCB	Change Control Board
CCFF	Crew Compartment Fit & Function
CDDT	Count Down Demonstration Test
CDH	Constant Delta Height
CDR	Commander
CDU	Coupling Data Unit
CM	Command Module
CMC	Command Module Computer
CMP	Command Module Pilot
CMPS	Command Module Pilot Simulator
CMS	Command Module Simulator
CO_2	Carbon Dioxide
COAS	Crewman Optical Alignment Sight
COMM	Communications
CSI	Concentric Sequence Initiation
CSM	Command/Service Module
CWG	Constant Wear Garment
DAP	Digital Auto-pilot
DCPS	Data Collection Platform Simulator
DEDA	Data Entry and Display Assembly
DELTA-H	Height Difference
DELTA-M	Momentum Change
DELTA-P	Pressure Difference
DELTA-V	Velocity Change
DELTA-R	Change of Radius Vector
DELTA-V_C	Change of Orbital Velocity
DOI	Descent Orbit Insertion
DPS	Descent Propulsion System
DSKY	Display & Keyboard
DTO	Detailed Test Objective
EASEP	Early Apollo Surface Experiment Package
ECO	Engine Cut off
ECS	Environmental Control System
ED	Explosive Device
EDS	Emergency Detection System
EECOM	Emergency Environmental & Consumables
EI	Entry Interface
ELS	Earth Landing System
EMS	Entry Monitor System
EMU	Extra-vehicular Mobility Unit (suit and backpack)
EVA	Extra-vehicular Activity
FMES	Full Mission Engineering Simulator
FCOD	Flight Crew Operations Division
FOD	Flight Operations Division
G	Force exerted by gravity or by acceleration or deceleration
G&C	Guidance & Control
G&N	Guidance & Navigation
GDC	Gyro Display Coupler

GET	Ground Elapsed Time
GFE	Government Furnished Equipment
GSOPS	Guidance and Navigation System Operations Plans
H-Dot	Time Derivative of height (Altitude), Descent rate or Ascent Rate
ICDU	Inertial Coupling Data Unit
IGM	Interactive Guidance Mode
IMU	Inertial Measurement Unit
ISA	Interim Storage Assembly
I_{sp}	Specific Impulse
ISS	Inertial Sensor System
IU	Instrument Unit
IVT	Intra-vehicular Transfer
KC-135	Training aircraft for low gravity simulations
L&A	Landing & Approach
LAM	Landing Area Map
LCG	Liquid Cooled Garment
LEB	Lower Equipment Bay
LEC	Lunar Equipment Conveyor
LET	Launch escape tower
LEVA	Lunar Extravehicular Visor Assembly
LGC	Lunar Module Guidance Computer
LiOH	Lithium Hydroxide air scrubber
LLRF	Lunar Landing Research Facility
LLTV	Lunar Landing Training Vehicle
LM	Lunar Module
LMP	Lunar Module Pilot
LMPS	Lunar Module Pilot Simulator
LMS	Lunar Module Simulator
LOI	Lunar Orbit Insertion
LOS	Loss of Signal
LPD	Landing Point Designator
LRL	Lunar Receiving Laboratory
LV	Low voltage
MCC	Mid Course Correction or Mission Control Center Houston
MESA	Modular Equipment Stowage Assembly
MIT	Massachusetts Institute Of Technology
MPAD	Mission Planning & Analysis Division
MQF	Mobile Quarantine Facility
MSC	Manned Space flight Center
MSFN	Manned Space Flight Network
ORDEAL	Orbital Rate Drive Electronics for Apollo and LM
O_2	Oxygen
OPS	Oxygen Purge System
OX	Oxidizer
PAD	Pre-advisory Data
PAO	Public Affairs Office
PDI	Powered descent initiation
PGA	Pressure Garment Assembly
PGNS	Primary Guidance and Navigation Subsystem
PLSS	Portable Life Support System
PM	Phase modulation
POGO	Up and down oscillations/ POGO Suppression System
PSE	Passive Seismic Experiment
PTC	Passive Thermal Control
PUGS	Propellant Utilization & Gauging System
PU	Propellant Unit
PYRO	Pyrotechnic device
REFSMMAT	Reference To Stable Member Matrix
RCS	Reaction Control System

RCU	Remote Control Unit	**VERBS**	
RDS	Rocketdyne Digital Simulator	1	Display Octal Data
R-Dot	Rate of change in range, rate of	6	Display Decimal Data
	approach	16	Monitor Decimal Data at half second
REV	Revolution, one orbit		intervals
S-IC	Saturn first stage	32	Recycle
S-IIC	Saturn second stage	37	Change Program
S-IVB	Saturn third stage	49	Crew Defined Maneuver
SCS	Stabilization & Control System	64	Calculate Orbital Parameters
SESL	Space Environmental Simulation	77	Set Rate Command/ Attitude Hold
	Laboratory		Mode in Digital Auto Pilot
SEP	Separation	82	Request Orbit Parameter Display
SIM	Simulator or Simulation	83	Request Rendezvous Parameter Display
SM	Service Module	87 & 88	Verb 87 and Verb 88 were used as a
SPS	Service Propulsion System		toggle by Collins to alternately lock out
SRC	Sample Return Container		and release the VHF Ranging Data in
STABLE I	Capsule floating nose-up		and out of the computer
STABLE II	Capsule floating on its side	89	Start Rendezvous Final Attitude
TEPHEM	Time of Ephemeris		Maneuver
TIG	Time of Ignition	90	Request Rendezvous Out Of Plane
TLI	Trans-lunar Injection		Display
TM	Telemetry	95	Inhibit State Vector Update (via
TPI	Terminal Phase Initiation		Navigation)
TTCA	Thrust Translation Controller Assembly		
UCD	Urine Collection Device	**PROGRAMS**	
V_I	Inertial Velocity	00	Computer Idling
WIF	Water Immersion Facility	12	LM Powered Ascent Guidance
Y-dot	Time derivative of pitch	20	Rendezvous Navigation
W-Matrix	Position Error (ft), Velocity Error (fps),	21	Ground Track Determination
	Radar Bias Angle Error (milliradians)	22	Rendezvous Radar Lunar Surface
			Navigation
		23	Star/Landmark Navigation Measurement
NOUNS		30	External Delta-V
4	Gravity Error Angle	34	Terminal/Transfer Phase Initiation
20	ICDU Angles	39	Proposed new program by Mike Collins
22	New Angles	40	DPS Thrusting
33	Time Of Ignition	41	RCS Thrusting
34	Time Of Event	47	Thrust Monitor
42	Apogee and Perigee	51	IMU Orientation Determination
44	Apogee and Perigee Altitude	52	Platform Alignment
49	Delta-R, Delta-V	57	Lunar Surface Alignment
65	Sampled LGC Time	63	LM Descent Landing Maneuver Braking
76	Desired Horizontal, Radial Velocity	64	LM Descent Landing Maneuver
	Cross Range Distance		Approach
77	Delta-T to Engine Cut-Off	68	Landing Confirm
85	Velocity to be Gained (VG)X (up),	76	Target Delta-V
	VGY(Rt), VGZ(Fwd) in fps		
86	Delta V in X, Y, Z